INTRODUCTION TO THE THERMODYNAMICS OF BIOLOGICAL PROCESSES

Prentice Hall Advanced Reference Series

Physical and Life Sciences

AGNEW AND MCCREERY, EDS. *Neural Prostheses: Fundamental Studies*
BALLENTINE *Quantum Mechanics*
BINKLEY *The Pineal: Endocrine and Nonendocrine Function*
BLACKWELL *Poisonous and Medicinal Plants*
CAROZZI *Carbonate Rock Depositional Models: A Microfacies Approach*
CHAFETZ *Nutrition and Neurotransmitters: The Nutrient Bases of Behavior*
EISEN *Mathematical Methods and Models in the Biological Sciences: Linear and One-Dimensional Theory*
EISEN *Mathematical Methods and Models in the Biological Sciences: Nonlinear and Multidimensional Theory*
FRASER *Clastic Depositional Sequences: Process of Evolution and Principles of Interpretation*
HAZELWOOD *The Endocrine Pancreas*
JEGER, ED. *Spatial Components of Plant Disease Epidemics*
LIBOFF *Kinetic Theory: Classical, Quantum, and Relativistic Descriptions*
MCLENNAN *Introduction to Nonequilibrium Statistical Mechanics*
MEFFE AND SNELSON *Ecology and Evolution of Livebearing Fishes (Poeciliidae)*
PLISCHKE AND BERGERSEN *Equilibrium Statistical Physics*
SORBJAN *Structure of the Atmosphere Boundary Layer*
VALENZUELA AND MYERS *Adsorption Equilibrium Data Handbook*
WALDEN *Genetic Transformation in Plants*
WARREN *Evaporite Sedimentology: Importance in Hydrocarbon Accumulation*
WARD *Fermentation Biotechnology*
WONG *Introductory Nuclear Physics*

PRENTICE HALL
Biophysics and Bioengineering Series
Abraham Noordergraaf, Editor

AGNEW AND MCCREERY, EDS. *Neural Prostheses: Fundamental Studies*
JOU AND LLEBOT *Introduction to the Thermodynamics of Biological Processes*

FORTHCOMING BOOKS IN THIS SERIES (*tentative titles*)

ALPEN *Radiation Biophysics*
COLEMAN *Integrative Human Physiology: A Quantitative View of Homeostasis*
DAWSON *Engineering Design of the Cardiovascular System*
GANDHI, ED. *Biological Effects and Medical Applications of Electromagnetic Energy*
HUANG *Principles of Biomedical Image Processing*
MAYROVITZ *Analysis of Microcirculation*
SCHERER *Respiratory Fluid Mechanics*
VAIDHYANATHAN *Regulation and Control in Biology*
WAAG *Theory and Measurement of Ultrasound Scattering . . .*

INTRODUCTION TO THE THERMODYNAMICS OF BIOLOGICAL PROCESSES

D. Jou
J. E. Llebot

Translated from the Spanish language by
Cathy Flick
John Woolman Enterprises
302 S. W. 5th Street
Richmond, Indiana 47374

Prentice Hall
Englewood Cliffs, New Jersey 07632

Library of Congress Cataloging-in-Publication Data

Jou, D. (David)
[Introducción a la termodinámica de procesos biológicos. English]
Introduction to the thermodynamics of biological processes / D.
Jou, J. E. Llebot ; translated from the Spanish language by Cathy
Flick.
p. cm. — (Prentice Hall advanced reference series)
(Prentice Hall biophysics and bioengineering series)
Translation of: Introducción a la termodinámica de procesos
biológicos.
Includes bibliographical references.
ISBN 0-13-502881-7
1. Nonequilibrium thermodynamics. 2. Biochemistry. I. Llebot,
J. E. (Joseph Enric) II. Title. III. Series: Prentice Hall
biophysics and bioengineering.
QP517.T48J6813 1990
574.19—dc20 89-72189
CIP

Editorial/production supervision: Raeia Maes
Cover design: Ben Santora
Manufacturing buyer: Denise Duggan

Prentice Hall Advanced Reference Series

Prentice Hall Biophysics and Bioengineering Series

Printed in the United States of America

10 9 8 7 6 5 4 3 2 1

ISBN 0-13-502881-7

Prentice-Hall International (UK) Limited, *London*
Prentice-Hall of Australia Pty. Limited, *Sydney*
Prentice-Hall Canada Inc., *Toronto*
Prentice-Hall Hispanoamericana, S.A., *Mexico*
Prentice-Hall of India Private Limited, *New Delhi*
Prentice-Hall of Japan, Inc., *Tokyo*
Simon & Schuster Asia Pte. Ltd., *Singapore*
Editora Prentice-Hall do Brasil, Ltda., *Rio de Janeiro*

Contents

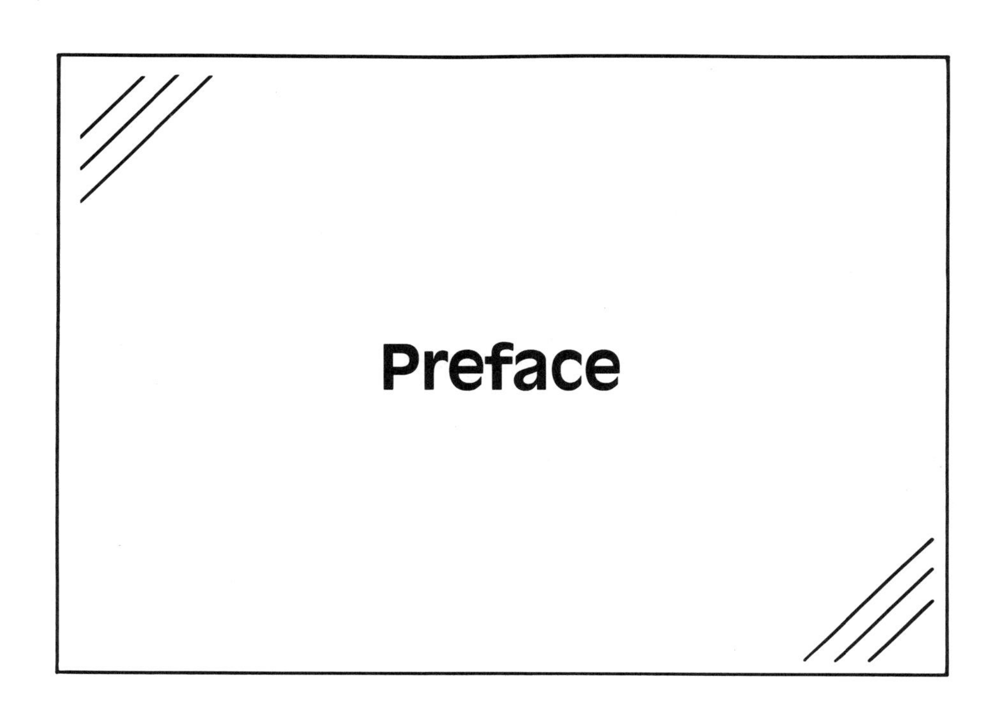

Preface

The goal of this book is to provide an introduction to nonequilibrium thermodynamics and its applications in biological systems. In biological systems, equilibrium means death. Therefore, mastery of the basic ideas that govern the nonequilibrium situation is of primary importance in order to understand biological systems, which are found outside equilibrium even in their simplest states (resting states). Even though in many cases there is no other alternative except to resort to detailed solutions of the equations of chemical kinetics and diffusion, thermodynamic ideas (macroscopic and globalizing in nature) help establish a working scheme capable of accommodating, with scope and clarity, the broad phenomenological outlines of biological processes.

While the thermodynamics of nonequilibrium systems has been treated in many texts from the point of view of physics, only a few include treatments related to questions of interest in biology. They also often use a formalism at too high a level for nonspecialists in the subject. In this book, we propose to establish the fundamentals of nonequilibrium thermodynamics in terms of a simpler formalism, in such a way that the text will be accessible to science students and researchers in any specialty. We have chosen applications that deal with biologically relevant processes, in order to illustrate what kind of information nonequilibrium thermodynamics has to offer in the study of problems of biological interest.

The text can be divided into two sections, which deal respectively with the linear and the nonlinear theory. The linear theory studies the restrictions imposed by the second law of thermodynamics on the phenomenological equations that relate the fluxes (heat flux, diffusion flux, reaction rate) with the thermodynamic forces (temperature gradient, chemical potential gradient, affinity). A central role is played by the Onsager reciprocity relations in reducing the number of independent coefficients that appear in the phenomenological equations. Among other interesting results, this part establishes a thermodynamic limit for the efficiency of transfer of free energy between different processes (one a producer and the other a consumer of free energy), which illustrates the advantages of using nonequilibrium thermodynamics in the treatment of bioenergetic processes.

On the other hand, thermodynamics poses an apparent paradox in biology: How can physical evolution toward disorder be compatible with biological evolution toward order? The classical response has stressed that biological systems are not isolated systems. This allows but does not explain the spontaneous structurization processes that are so abundant in biology.

This problem is the focus of the second part of the book, in which we pick up nonlinear theory and its application to the spontaneous appearance of dissipative structures in systems far from equilibrium (spatial structures, temporal rhythms). These self-organizing aspects are especially important in the study of the structurization phenomena in living systems (morphogenesis, population dynamics, biological clocks).

In the final chapters of the book, we focus our attention on a topic of great conceptual interest which has monopolized the attention of many scientists in recent years: deterministic chaos and fractal geometries. We will see how a simple deterministic equation can lead to chaotic behavior, without temporal regularity and very sensitive to the initial conditions. This has led to a true scientific revolution, whose importance has been compared by many authors with that of the great relativistic and quantum revolutions of the first quarter of the century.

Unlike the existing literature to date on these topics, our text is directed toward students and researchers especially interested in the interrelation between physics and biology who lack previous training in nonequilibrium thermodynamics. The book therefore offers a concise and direct introduction to the basic ideas of this science, and extends to the most influential research areas of recent years.

David Jou
Josep Enric Llebot

Acknowledgments

We thank our colleagues at the Universitat Autònoma de Barcelona, and especially Professor J. Casas-Vázquez, for an exciting research and teaching environment, without the stimulation of which this book would not even have been begun.

We thank Editorial Labor for permission to reproduce Fig. 10.10 from the text *Biología celular y molecular* [*Molecular Cell Biology*] by Darnell, Lodish, and Baltimore.*

We also thank Prensa Científica for the authorization to reproduce the following illustrations:

Fig. 4.2: *Investigación y Ciencia*, April 1985, p. 37.
Fig. 10.1: *Investigación y Ciencia*, September 1980, p. 62.
Fig. 10.2: *Investigación y Ciencia*, September 1980, p. 57.
Fig. 10.4: *Investigación y Ciencia*, September 1980, p. 64.
Fig. 10.7: *Investigación y Ciencia*, September 1978, p. 111.
Fig. 10.8: *Investigación y Ciencia*, February 1985, p. 92.
Fig. 11.2: *Investigación y Ciencia*, February 1987, p. 22.
Fig. 11.3: *Investigación y Ciencia*, September 1987, p. 23.
Fig. 11.5: *Investigación y Ciencia*, April 1985, p. 37.

---

* J. E. Darnell, H. F. Lodish, and D. Baltimore, *Molecular Cell Biology*. New York: Scientific American Books, 1986.

INTRODUCTION TO THE THERMODYNAMICS OF BIOLOGICAL PROCESSES

LINEAR THEORY

1

Brief Review of Equilibrium Thermodynamics

1.1 FIRST LAW OF THERMODYNAMICS. CONSERVATION OF ENERGY

The first law of thermodynamics identifies heat as a form of energy. It took a long time for this idea, which seems elementary to us today, to make any headway. The concept was not formulated until the 1840s, mainly as a result of the investigations of Mayer and Joule. Before this, it was thought that heat was an indestructible, weightless substance (caloric) that had nothing to do with energy. This point of view, formulated by Lavoisier, played an important role in the scientific revolution that gave rise to chemistry in the eighteenth century. The caloric theory proved to be a productive one in its day. For example, it was helpful in the work of Carnot, which was so important in establishment of the second law, and also in the correct calculation of the speed of sound by Laplace.

Once the equivalence of energy and heat is realized, the first law postulates conservation of energy. We then note that the first law is not simply a formulation of this conservation principle, but also assumes an identification of the nature of heat.

In order to begin our work, we must first of all describe the object of our study. Let us call that part of the universe under consideration *the system*. We customarily speak of isolated, closed, or open systems. A system is isolated when it does not exchange energy or mass with its surroundings. If the system exchanges energy but not mass, it is closed. If the system exchanges mass and energy with the surroundings, we say the system is open. It is possible to conceive a system that exchanges mass but not energy, but the result is not very realistic. Naturally, the systems of the most interest, considering their applications in biology, are open systems, since most biological systems are open. But in order to begin the thermodynamic study of systems, it is convenient to start with the simplest (isolated systems) and then extend the formalism to the study of open systems. We also distinguish between systems in equilibrium states and in nonequilibrium states. Generally we associate the concept of equilibrium with absence of the exchange fluxes. This situation is often related to the spatial homogeneity of the parameters characterizing the system, which in equilibrium do not depend on time. A nonequilibrium state, on the other hand, is a state with net exchange of mass or energy, and its characteristic parameters in general depend on position and time. If they do not depend on the latter, they require participation of the surroundings to maintain their values (a nonequilibrium steady state).

In order to specify the system that we want to study, let us ask several questions: On which variables do we focus our attention? What relationships exist between these variables? What is the minimum number of essential variables for the unequivocal description of the system at a certain instant? This minimum set of variables constitutes the specification of what we call in thermodynamics *the state of the system.* In other words, the state of a system is described by the value of certain measurable variables. In a fluid system (a gas or a liquid), a state is defined first of all by the volume V; certainly this variable is not sufficient. In fact, for example, the fluid can be more or less hot and so it is necessary to introduce a new variable: the temperature (either the empirical temperature, which is the one we obtain from any thermometer, or else the absolute temperature T, defined according to the second law of thermodynamics). The definition and rigorous conceptualization of temperature is subtle and rather complicated. At this introductory level, the primitive concept of temperature is enough for us.

Experiment shows that all the other properties of a fluid system are a function of the volume V and the temperature T. Thus, in particular, the pressure (the force exerted by the fluid per unit area on the surface of the container) is a function

$$p = p(T, V)$$

called *the thermometric equation of state*; and the internal energy of each state U, which we define next, is a function

$$U = U(T, V)$$

which is called *the calorimetric equation of state.* Each one-component fluid is characterized by these two equations of state (calorimetric and thermometric).

We have introduced the concept of internal energy, a concept that is much less intuitive than the concept of pressure. How do we define and evaluate it from the thermodynamic point of view? Let us start from a reference state A characterized by the values V_A, U_A. Let us assume that the system is thermally isolated; that is, if we bring it near a heat source (such as a flame), its state does not change. We can do work on the system, for example by agitation. As in the famous experiment by Joule, the work done can be measured by the decrease in the potential energy of a weight, and can be given by

$$w = mg(h_i - h_f)$$

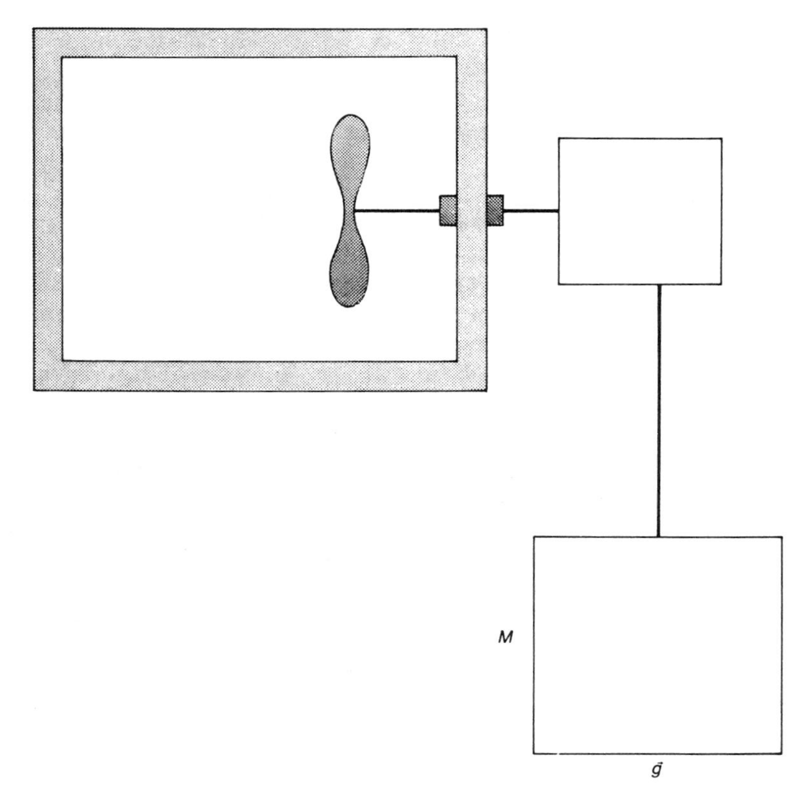

Figure 1.1 The Joule experiment. Upon lowering, the mass m (moved by gravity) gives up its potential energy to the thermodynamic system, whose energy increases.

The work w done on the system (work that increases its energy) in this case is equal to the decrease in the potential energy of the weight mg when it descends from an initial height h_i to a final height h_f. This process does not change the volume of the system but increases its temperature, as can be confirmed with a thermometer. Thus, if we assign an arbitrary value U_A to the internal energy of the system in the state A(V_A, T_A), its internal energy U_B in the final state B(V_A, T_B) will be necessarily $U_B = U_A + w$, that is, the initial energy plus the work done on the system.

Another way to do work on the system is to change its volume. In this case, according to the mechanical definition of work (work is equal to the scalar product of the force times the distance traveled), we have (Fig. 1.2)

$$dw = F \cdot dx = (ps) \cdot dx = p(sdx) = -pdV \tag{1.1}$$

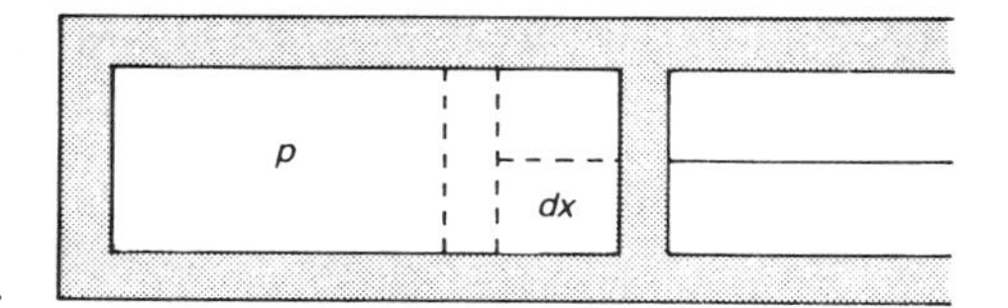

Figure 1.2 Work of compression.

In these equations, we consider first of all that $F = ps$ (force = pressure times surface area), and second that $sdx = -dV$ (surface area times height equals volume). If dx is positive, the system is compressed and the change dV in the volume is negative: the volume decreases in compression and increases in expansion. During the volume change, considering the work and the reference value U_A assigned to the initial state, we can assign the following energy value to the final state (V_B, T_B):

$$U_B = U_A + \int_A^B dw$$

where we integrate the expression for the work throughout the process, which takes the system from state A to state B.

By doing work on the system either by agitation or by changing the volume, we can arrive at a final state B starting from a given state A. From mechanics, we know how to measure the work w_{AB} done on the system in passing from state A to state B; then we define

$$U_B = U_A + w_{AB}$$

It is possible that we cannot pass from the reference state A to another given state B by means of this procedure. Thus, for example, during agitation the temperature of an isolated system can increase but it cannot decrease. In this case, we know from experience that if we can pass from state B to state A, we can define the internal energy of state B as

$$U_B = U_A - w_{BA}$$

where w_{BA} is the work done on the system in passing from B to A.

This procedure allows us to assign to each state of the system described by T, V an internal energy value $U(T, V)$. If the walls of the system are not thermally insulating,

$$U_B - U_A \neq w_{AB}$$

that is, the work done on the system in passing from state A to state B is not equal to the difference in the internal energies that have been assigned to each state of the system according to the theoretical procedure in the previous paragraph. In this case, we define the value of q_{AB}, the heat gained by the system in passing from A to B, as

$$q_{AB} = U_B - U_A - w_{AB} \tag{1.2}$$

This expression, which constitutes a definition of heat, is the first law of thermodynamics, which can be expressed in a more general form as

$$\Delta U = q + w \tag{1.3}$$

Some texts use another form of the first law with respect to the signs. The possible discrepancies come from the sign convention used with respect to w and q. In this text, we consider w and q positive if work is done on the system (it is compressed) and if it receives heat; and they are respectively negative in the opposite case. Sometimes (and from a less anthropocentric point of view) w is considered positive when it is the system that does the work, in which case the sign of w is negative in the first law.

Equations (1.2) and (1.3) define heat and allow us to measure it. If we assume that we also know how to measure the temperatures, then we can measure the specific heats and latent heats of a change in state, defined respectively as

$$dq = mcdT \tag{1.4}$$

$$dq = mc' \tag{1.5}$$

where m is the mass of the system, c is its specific heat, and c' is the latent heat of the phase change. In the first case, the heat that is transferred to the system (dq) increases its temperature by the increment dT. In the second case, that is, during a phase change (passing from solid to liquid, from liquid to vapor, and so on), the heat transferred to the system does not increase the temperature since the temperature remains constant while the phase change is occurring, but the heat is used to modify the internal structure of the particles that compose the system. The concept and some values of c and c' can be found in any elementary physics text, so we do not need to pay any more attention to them here.

1.2 SECOND LAW OF THERMODYNAMICS. IRREVERSIBILITY

The first law imposes a restriction on the evolution of systems. An isolated system can change only between states that have the same energy and not to states of greater or lesser energy. The first law does not establish any direction for the evolution of processes. This is in contrast to what we observe in nature, where preferred directions are apparent (heat is transmitted from the hot body to the cold body, matter is transferred from zones of higher concentration to zones of lower concentration, and so on). In order to establish the direction of the evolution of processes in accord with observations, we need to formulate a new law that is independent of the first law. This new law, which has various equivalent formulations, is *the second law of thermodynamics*. It was formulated for the first time in the year 1850 by two authors independently, the German Rudolf Clausius and the Englishman William Thomson (Lord Kelvin), in the following terms:

Clausius Formulation: *It is impossible to construct a device which transfers heat from a cold body to a hot body in a cyclic process without any other effect.*

Kelvin Formulation: *It is impossible to construct a device which can raise a body by extracting thermal energy from another body in a cyclic process and without any other effect.*

We will not rigorously study the equivalence of these two formulations, but rather we will restrict ourselves to briefly discussing them. Reflecting on the formulation of Clausius, we can see an apparent contradiction: refrigerators. They extract heat from the (cold) interior to the (hot) exterior in order to keep the interior temperature low. Nevertheless, this is not their only effect, since they also consume electricity. This false counterexample allows us to evaluate the role of the restriction *without any other effect* that appears in the Clausius formulation. In other words, heat does not pass spontaneously from a cold body to a hot body.

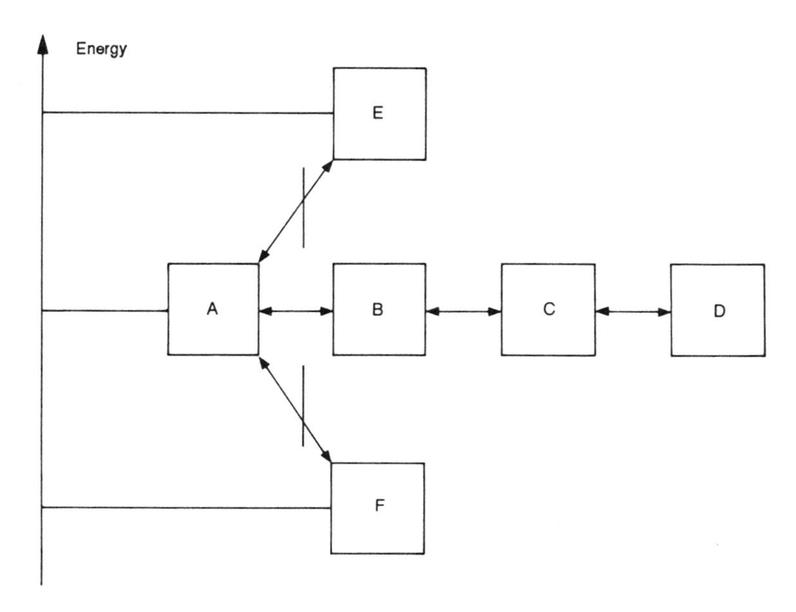

Figure 1.3 Energy of states A, B, C, D, E, F of a system.

The Kelvin formulation is also easily understood by means of counterexamples. One such counterexample involves a boat that moves over the sea without consuming fuel, extracting the necessary energy from the thermal energy of the sea by the operation of its motors. Despite the great economic attractiveness of this idea, since the thermal energy of the sea is enormous (virtually inexhaustible) and inexpensive, the Kelvin formulation forbids its realization: it is not possible to convert heat completely into work, although work can be completely converted into heat. Therefore, while according to the first law heat and work are equivalent forms of energy exchange, the second law radically changes their equivalence, since work can go completely to heat but heat cannot be completely transformed to work. The first principle prohibits or, better, simply states the impossibility of perpetual motion of the first kind, in which energy is extracted from nothing. The second principle states the impossibility of perpetual motion of the second kind, which (as discussed previously) is more subtle: it does not violate conservation of energy, but transforms heat completely into work.

The mathematical formulation of the second law (due to Clausius, 1865) introduces a new state function: the *entropy*, defined as

$$S_{\mathrm{B}} = S_{\mathrm{A}} + \int_{\mathrm{A}}^{\mathrm{B}} (dq)_{\mathrm{rev}}/T \tag{1.6}$$

where S_{A} is the (arbitrary) value that we assign to the entropy of the reference state A; T is the absolute temperature (or the temperature in kelvins, that is, T = Celsius temperature + 273.16°); dq_{rev} is the heat exchanged in an ideal reversible process. The second law asserts that in an isolated system it is possible to pass from a state A to a state B only if $S_{\mathrm{B}} \geq S_{\mathrm{A}}$, and that it is impossible to go in the opposite direction. When $S_{\mathrm{B}} = S_{\mathrm{A}}$, it is possible to go both from A to B and from B to A, and the process is called *reversible*.

In order to clarify the relation between this mathematical formulation and the verbal and intuitive formulation of Clausius, let us consider the following example:

A system is composed of two identical subsystems, one at $T_1 = 400$ K and the other at $T_2 = 200$ K. In this case, the two temperatures constitute the state variables of the system, since we assume that the volume of the subsystem does not change in spite of the change in temperature. We must calculate the entropy corresponding to the states B and C, defined as follows:

$$\text{state B } (T_1 = 300 \text{ K}, T_2 = 300 \text{ K})$$

$$\text{state C } (T_1 = 500 \text{ K and } T_2 = 100 \text{ K})$$

We have chosen the values in such a way that the sum of the absolute

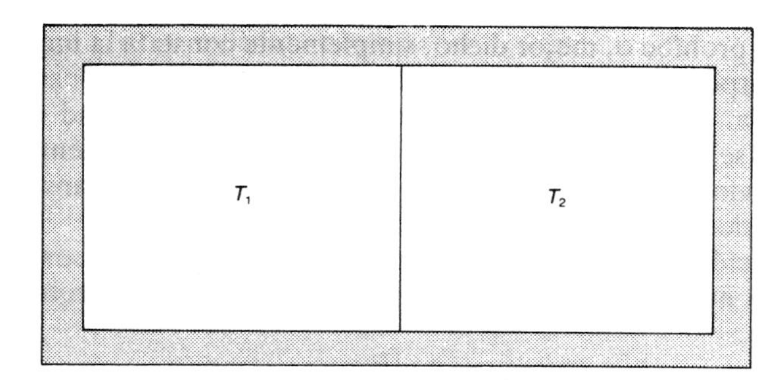

Figure 1.4 Two identical subsystems in mutual thermal contact and isolated from the outside world.

temperatures of each one of the two states is always the same; then, if we suppose that the internal energy is the specific heat multiplied by the temperatures, we guarantee (when dealing with two subsystems with equal properties) that passing from A to B and from A to C does not contradict the first law. This condition (constant sum of the temperatures) is by no means always definite, since if the two subsystems are made from different materials they have different specific heats, and therefore they do not satisfy the previous condition. In general, the first step in determining whether or not a process is spontaneous is to test whether it is possible according to the first law.

Once it has been ascertained that the process is compatible with the first law, we make use of the second law to test its possible spontaneity. The total entropy of the isolated system is the sum of the entropy of the two subsystems,

$$S_{tot} = S_1 + S_2$$

The variation in entropy of each subsystem, according to definition (1.6) and according to the definition of specific heat (1.4) is given by

$$\Delta S_1 = \int_i^f dq/T = \int_{T_{1i}}^{T_{1f}} mc_1 dT/T = mc_1 \ln(T_{1f}/T_{1i})$$

$$\Delta S_2 = \int_i^f dq/T = \int_{T_{2i}}^{T_{2f}} mc_2 dT/T = mc_2 \ln(T_{2f}/T_{2i})$$

where i and f symbolize the initial and final state respectively and T_{1f}, T_{2f}, T_{1i}, and T_{2i} designate the final and initial temperatures of each subsystem.

The entropy of state B, if we take state A as the reference, is

$$S(300, 300) = S(400, 200) + mc_2 \ln(300/200) + mc_1 \ln(300/400)$$

or, equivalently,

$$S(300, 300) = S(400, 200) + 0.12mc_1$$

since $c_1 = c_2$ and $T_{1i} = 400$ K, $T_{1f} = 300$ K, $T_{2i} = 200$ K, $T_{2f} = 300$ K.

Analogously, the entropy of state C is

$$S(500, 100) = S(400, 200) + mc_2 \ln(100/200) + mc_1 \ln(500/400)$$

$$S(500, 100) = S(400, 200) - 0.47mc_1$$

With this example, we have illustrated how we calculate entropies: we use the expression $dq = mcdT$ for the heat and integrate the resulting expression from the initial state to the final state of each subsystem. At this point, we omit the subtleties which arise in considering whether or not the process is reversible (which we will soon discuss) and we assume that heat transfer from each subsystem is very slow and is achieved in contact with a hypothetical heat source at the same temperature. Once we have evaluated the entropy, we can ask which of these two processes (from A to B or from A to C) will be spontaneous and which will not be spontaneous. Since the global system is isolated, we have

$$S_B - S_A = 0.12mc_1 > 0 \qquad \text{spontaneous process}$$

$$S_C - S_A = -0.47mc_1 < 0 \qquad \text{nonspontaneous process}$$

We have thus verified that the mathematical criterion that we have used as a formulation of the second law is in perfect accord with our observations: the process from A to B, in which the temperatures tend to equalize by passing heat from the hot body to the cold body, is spontaneous. On the other hand, the process from A to C, in which heat flows from the cold body to the hot body and the temperatures even "unequilibrate," is not spontaneous.

Finally, recall that in referring to spontaneous processes we have not had any time scale in mind. There are processes allowed in principle by the second law which takes place very slowly and can take thousands of years to complete. Classical thermodynamics does not include time in its considerations, since the slowness or rapidity of a process is irrelevant to the second law. On the scale of our lifetimes, nevertheless, we tend to consider as spontaneous only those processes which are sufficiently rapid so that their realization is perceptible. Here we use the concept of spontaneous process in its more general sense, independently of its evolutionary time scale.

1.3 EFFICIENCY OF HEAT ENGINES

A traditional and especially illustrative application of the second law is calculation of the maximum efficiency of heat engines, a problem that was treated for the first time by Sadi Carnot (1824). Let us imagine the following situation (Fig. 1.5): a heat source (a large heat reservoir) at the absolute temperature T_1 transfers a quantity of heat q_1 to the engine R. This engine, during one cycle, transforms one part of q_1 into work w; and the other part q_2 is transferred as heat to the reservoir at the lower absolute temperature T_2. In this process, part of the heat is transformed into work, in accord with the objective of heat engines. We observe that, according to Kelvin formulation, we cannot transform heat completely into work. Therefore, it is necessary that a certain quantity of heat q_2 be transferred to a heat reservoir at the lower temperature T_2. Let us now apply the first law. In one cycle, the internal energy of the engine R does not change. Therefore,

$$q_1 - q_2 = w$$

that is, the heat received q_1 minus the heat transferred q_2 is equal to the work produced w.

The second law imposes the condition

$$-(q_1/T_1) + (q_2/T_2) \geq 0$$

where the equals sign applies in the case of a reversible process, and the inequality holds in all other cases. This expression has been obtained bearing in mind that the total change in entropy is the sum of the changes in the entropies of the heat source 1, the engine R, and the heat source 2.

$$\Delta S = \Delta S_1 + \Delta S_{\text{R}} + \Delta S_2$$

Now then, in one cycle the engine R returns to the same state from which it started, and therefore

$$\Delta S_{\text{R}} = 0$$

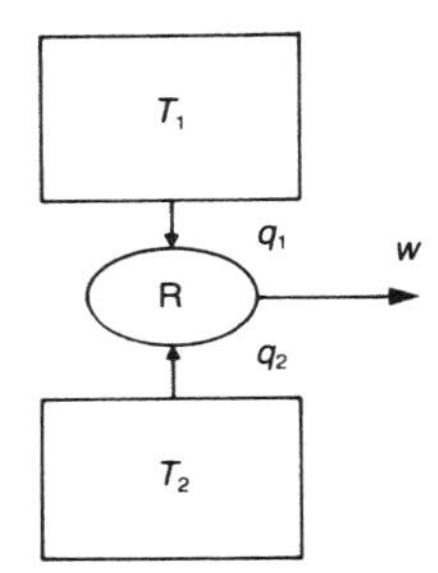

Figure 1.5 Heat engine. The heat source at temperature T_1 transfers a certain quantity of heat q_1 to the heat engine R, which transforms part of it into work w and transfers the rest to the heat source at temperature T_2 in the form of heat q_2.

since the final entropy coincides with the initial entropy. The change in the entropies of the heat sources (which are maintained at constant T_1 and T_2, since they are large enough so that the absorption or transfer of the heat does not produce appreciable changes in their temperatures) is simply the quotient of the heat exchanged divided by the absolute temperature. That is,

$$\Delta S_1 = -q_1/T_1 \qquad \Delta S_2 = q_2/T_2$$

where the minus sign in front of q_1 indicates that the source 1 has lost heat q_1 and that therefore its entropy has decreased.

If we define the efficiency of the engine as $\eta = w/q_1$, that is, as the fraction of heat which is transformed into work relative to the total heat received, we have

$$\eta = w/q_1 = (q_1 - q_2)/q_1 = 1 - (q_2/q_1)$$

and, according to the first and the second law,

$$\eta = w/q_1 = 1 - (q_2/q_1) \leq 1 - (T_2/T_1) \tag{1.7}$$

since $T_2/T_1 \leq q_2/q_1$. The expression (1.7), obtained by Carnot, indicates that the maximum efficiency of a heat engine is achieved when the engine operates reversibly (which is when the inequality is changed to an equality) and is determined by the temperatures of the heat sources between which the engine operates. This result is of great industrial interest, since it provides information about how to improve the efficiencies of engines and processes with heat exchange sources, and allows us to avoid futile efforts which we might make if we ignored this rule and tried to improve the real efficiency despite this maximum theoretical limit. On the other hand, this expression is not of great interest in biology, since most biological processes are virtually isothermal. The same expression permits a formulation of greater usefulness from the economic point of view,

$$\eta = \alpha w/\beta q_1$$

where α is the price at which the unit of work is sold and β is the price at which the unit of heat is bought, or the price of fuel.

In Chapter 6, we study a different formulation, which is much more appropriate for studying the efficiency of energy conversion processes and in particular bioenergetic processes, based on the thermodynamics of irreversible processes.

1.4 THE SECOND LAW FOR NONISOLATED SYSTEMS

Let us now suppose that the system is not isolated, that is, that it can exchange heat and work (but not matter) with the external medium. Previously, we insisted that the formulation of the second law is valid only for isolated systems. How can we reformulate it for the closed systems that we consider in this section?

Let us assume, for example, that the system is in contact with the atmosphere, which is maintained at a temperature T and at a pressure p (fixed). The second law can be formulated considering the combination of the surroundings plus the studied system as a large isolated system which we call the universe. In this case, we have

$$S_{\text{universe}} = S_{\text{system}} + S_{\text{surroundings}} \geq 0 \qquad (1.8)$$

However, this formulation is inconvenient, since it involves not only the system that is the object of our study, but also the surroundings. We need to find a formulation that is a function of only the variables of the system. This can be done if we bear in mind, on the one hand, the first law of thermodynamics, and on the other hand the definition of entropy. According to the first law (1.3), we have

$$\Delta U_{\text{system}} = q_s + w_s$$

and according to expression (1.1) for the work done on the system

$$w_s = -p\Delta V_{\text{system}}$$

we can determine the heat q_a gained by the surroundings based on the heat q_s gained by the system, since the two quantities are related according to

$$q_a = -q_s$$

Therefore

$$q_a = -q_s = -(\Delta U_s - w_s) = -(\Delta U_s + p\Delta V_s)$$

Since the ambient temperature is constant, the system both receives and transfers heat; given the definition of entropy in (1.6), we have

$$S_{\text{surroundings}} \equiv \Delta S_a = q_a/T = -(\Delta U_s + p\Delta V_s)/T$$

where we have used the previous formula for q_a. We can write the criterion (1.8) of the second law, using the expression for ΔS_a that we have just obtained, as

$$\Delta S_s + \Delta S_a = \Delta S_s - (\Delta U_s + p\Delta V_s)/T \geq 0$$

or in a more compact form

$$T\Delta S_s - \Delta U_s - p\Delta V_s \geq 0 \qquad (1.9)$$

Therefore, only processes that satisfy condition (1.9) will be spontaneous, which as we have seen is completely equivalent to the second law in its expression (1.8), but with the advantage that now it involves only the variables of the system proper; T and p, we recall, are common to the system and to the surroundings, and reflect the conditions under which the experiment is performed.

In order to formulate this condition in a more concise form, we define a new function called the *Gibbs function*, or the *Gibbs free energy* G, as

$$G = U + pV - TS \tag{1.10}$$

As a function of G, criterion (1.9) can be written as

$$(\Delta G_s)_{T,p} \leq 0 \tag{1.11}$$

where the subscripts T and p indicate that the change in G must be realized at constant temperature and pressure. This is the expression of the second law for closed systems subject to fixed temperature and pressure.

We can immediately pose other similar questions. For example, what would be the criterion for knowing if a process is spontaneous in a system with fixed T and V? The same expression (1.9) allows us to answer this question: if V is fixed,

$$\Delta V_s = 0$$

and therefore (1.9) is reduced to

$$T\Delta S_s - \Delta U_s \geq 0$$

This relation can be written in a more elegant form if we define the *Helmholtz function* or the *Helmholtz free energy* F as

$$F = U - TS \tag{1.12}$$

As a function of F, the previous criterion can be written as

$$(\Delta F_s)_{T,V} \leq 0 \tag{1.13}$$

where the subscripts T and V indicate that the temperature and the volume are held constant. This relation summarizes in a single formula the two important criteria for the evolution of systems: on the one hand, the systems with constant entropy usually considered by mechanics tend toward minimum energy; on the other hand, isolated systems tend toward maximum entropy. With the free energy F, maximization of entropy and minimization of energy are subtly interrelated. Thus, in a system with constant temperature and volume, only processes in which F decreases can occur spontaneously. At very low temperatures, the factor $T\Delta S$ associated with the changes in entropy is negligible compared with the term involving the change in energy ΔU, and the system tends in fact to minimize the energy. At elevated temperatures, the entropy term predominates and the system appears to maximize the entropy.

In fact, relation (1.9) allows us to formulate the expression of the second law for an isolated system in which the internal energy and the volume remain constant, and for a system with constant entropy and volume (the latter is the case for usual mechanical systems, if we ignore dissipation and its thermal effects).

In the case of an isolated system (U and V constant), we have

$$\Delta U_s = \Delta V_s = 0$$

and (1.9) reduces to

$$T(\Delta S_s)_{U,V} \geq 0 \tag{1.14}$$

Since the absolute temperature T is always positive, we recover the previous result, according to which only processes that increase the entropy are possible in an isolated system.

In the case of a mechanical system with constant entropy and volume, we have

$$\Delta S_s = \Delta V_s = 0$$

and (1.9) is reduced to

$$-(\Delta U_s)_{S,V} \geq 0$$

For these conditions, the system tends toward the minimum energy; that is, only those processes for which the associated change in internal energy is equal to or less than zero are spontaneous.

Finally, a system with constant entropy and pressure ($\Delta p = 0$, $\Delta S_s = 0$) can spontaneously undergo those processes which satisfy the condition obtained starting from (1.9)

$$-(\Delta U_s) - p(\Delta V_s) \geq 0 \tag{1.15}$$

This condition is written as a function of the *enthalpy* H, defined as

$$H = U + pV \tag{1.16}$$

if

$$(\Delta H_s)_{p,S} \leq 0 \tag{1.17}$$

in which again the subscripts p and S indicate that both variables are held constant.

In summary, we must remember that the second law does not state that the entropy must always increase. This is definitely the case only for isolated systems. For closed systems that are not isolated and are subject to various conditions, the second law can be defined in terms of the decrease in the Gibbs free energy G (constant T and p), the decrease in the Helmholtz free energy F (constant T and V), the decrease in the enthalpy H (constant S and p), or the decrease in the internal energy U (constant S and V).

1.5 MICROSCOPIC INTERPRETATION OF ENTROPY

The Austrian physicist Ludwig Boltzmann proposed in 1872 a microscopic interpretation of entropy, that is, an interpretation of this concept in terms of the properties of molecular movement. Up to now, all our assertions have ignored the nature of the systems: These laws are as valid for continuous systems as for discrete systems composed of small particles (atoms and molecules). At the time when Boltzmann formulated his interpretation, the concept of molecules was only a working hypothesis, rejected by most scientists of the period. We need to bear this in mind in order to realize how far ahead of his time this brilliant scientist was.

Boltzmann related the entropy S of any state of a system to the number W of microstates compatible with the macrostate, according to the relation

$$S = k \ln W \tag{1.18}$$

where k, the Boltzmann constant, has the value $k = 1.38 \cdot 10^{-23}\ \mathrm{J \cdot K^{-1}}$ (J means joules, K means kelvins). This is a universal constant; it does not depend on the system or on any specific circumstances, and this is the reason for its great importance.

In order to understand what a microstate is, let us make use of a very simple example. Let us consider that the system we want to study is a gas formed by only four molecules, which can be located indistinguishably in either of two containers in contact. The macrostates (states that are studied by thermodynamics, which can be sensed from the macroscopic world—that is, with the naked eye) can be described by two numbers, N_a and N_b, which are the number of gas molecules in container a and in container b respectively. Next, let us list the possible values of N_a and N_b.

Macrostate	
N_a	N_b
0	4
1	3
2	2
3	1
4	0

Therefore, there are five distinct macrostates accessible to the system, each one characterized by the specification of the number of molecules in each container (however, without specifying precisely which particular

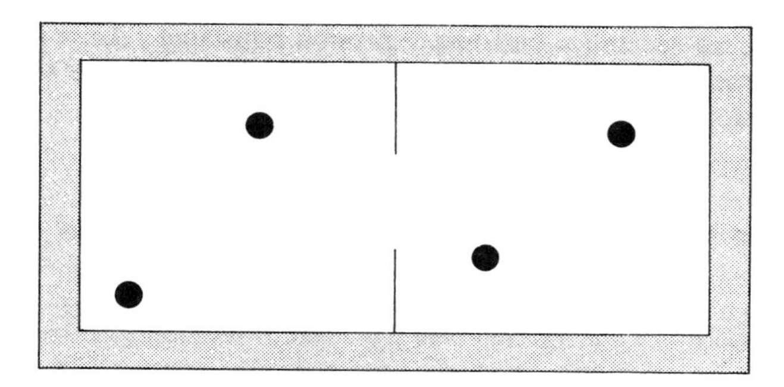

Figure 1.6 Two containers connected by a hole through which molecules can pass.

molecule occupies one or the other container). Knowing which particular microstate is occupied by the system at a given instant requires more detailed information, since we would need to know explicitly which molecules are in each part of the system. On the other hand, different distinct microstates can correspond to the same macrostate. Thanks to elementary combinatorics, it is easy to calculate the number of microstates that correspond to a given macrostate. In fact, the number of microstates W which corresponds to a definite distribution of molecules N_a, N_b in the two containers is expressed as

$$W_{N_a,\,N_b} = \binom{4}{N_a} = \binom{4}{N_b} = (N_a + N_b)!/N_a!N_b! \tag{1.19}$$

where the notation $N!$ is equivalent to $N(N - 1)(N - 2) \ldots (3)(2)(1)$. In order to be more explicit, let us consider the case of the macrostate ($N_a = 1$, $N_b = 3$). How many microstates correspond to it? If $N_a = 1$, there is only one molecule in container a. Now then, there are four possibilities for choosing this molecule, since a total of four molecules is available. Therefore, the number of microstates is calculated as

$$W_{1,3} = \binom{4}{1} = 4!/3!1! = 4$$

If $N_a = 2$ and $N_b = 2$, there are six possible choices for the two molecules in container a. Explicitly, the two molecules of a can be 1 and 2, or 1 and 3, or 1 and 4, or 1 and 3, or 2 and 4, or 3 and 4. Consequently, the number of microstates is six.

Thus, the number of microstates corresponding to each macrostate is as shown in the table

Macrostate		Number of microstates
N_a	N_b	W
0	4	1
1	3	4
2	2	6
3	1	4
4	0	1

From experience, we know that if at the beginning we have four molecules (or four moles) in container b and we open the conduit that communicates with container a, the final equilibrium state (in which the system remains stationary) will be the one in which the number of moles is the same in both containers—that is, it will correspond to a homogeneous distribution. This is precisely the macrostate that is compatible with the greatest number of microstates. Since the entropy increases when the number of microstates increases, we observe that the system tends (when it is isolated) toward the maximum entropy, toward the most probable state.

According to Boltzmann's interpretation, nature tends toward disorder, if we interpret the number of accessible microstates as disorder. In real situations, we do not focus only on the position of the molecules but also on their velocities; this must be kept in mind when the number of accessible microstates is exactly evaluated. The Boltzmann relation, most importantly, is the fundamental bridge between the macroscopic and microscopic worlds, and allows us to calculate (with the techniques of statistical mechanics) the properties of macroscopic systems composed of many particles, based on the theorems of mechanics and electrostatics/electrodynamics concerning the behavior of particles.

A well-known expression for the entropy may be obtained from (1.18) and (1.19) for large values of N_a and N_b. For large N, we may use the Stirling's approximation

$$\log N! = N \log N - N \tag{1.20}$$

to evaluate $\log N!$. Thus, from (1.18) and (1.19) we have

$$S = k \log \frac{(N_a + N_b)!}{N_a!\, N_b!} = k[\log (N_a + N_b)! - \log N_a! - \log N_b!] \tag{1.21}$$

Making use of (1.20), we find that this yields

$$S = -k\left[N_a \log \frac{N_a}{N_a + N_b} + N_b \log \frac{N_b}{N_a + N_b}\right] \tag{1.22}$$

We may write $p_a = N_a/(N_a + N_b)$ and $p_b = N_b/(N_a + N_b)$, where p_a and p_b obviously stand for the probability to find a particle in a or in b respectively. In this notation, (1.22) may be written as

$$S = -kN[p_a \log p_a + p_b \log p_b] \tag{1.23}$$

In the situation of the simple example used in this section, each particle has two possible states, a or b. The microstate of the total system is defined by which particles are in either state. In more complex situations, each particle may be found in one of M different states, defined not only through position in space but also in terms of the velocity of the particle. In such a general case, an immediate generalization of (1.23) is

$$S = -kN \sum_{i=1}^{M} p_i \log p_i \tag{1.24}$$

with p_i the probability of finding a given particle in state i. This expression for the entropy is quite well-known and, in fact, it was the one that Boltzmann used in his works on kinetic theory of gases. Note, however, that (1.24) is valid only when every particle behaves independently of all the other particles in the system, whereas definition (1.18) is more general, since it may be used in systems of interacting particles.

An expression similar to (1.24) was used by Shannon to evaluate the information in a message. The average information in a message with M possible alternatives is

$$H = -\sum_{i=1}^{M} p_i \log_2 p_i \tag{1.25}$$

where p_i is the prior probability of the ith alternative. Whereas the entropy in (1.24) is measured in units of energy/(absolute temperature), the information in (1.25) is expressed in bits. Formula (1.25), which is widely used in communication engineering and in digital computation, has also been used to define, in a quantitative way, the diversity of ecosystems, and to discuss sensory biophysics, molecular biology, and genetics.

1.6 THE SECOND LAW IN BIOLOGY

The transition from the microscopic world to the macroscopic world is full of perplexities. One of these presents itself when we consider the second law in biological systems. The paradox that is raised is the following: according to physics, according to thermodynamics, nature tends toward disorder, toward homogeneity. On the other hand, biological systems tend toward order and toward structurization. Do biological systems obey the laws of thermodynamics?

This paradox is only apparent. In the first place, biological systems are not isolated systems, but exchange energy and matter with the exterior world: they feed, they breathe, they excrete. A biological system dies soon after it becomes isolated. This demonstrates that the version or formulation of the second law that should be applied to biological systems is not the one for isolated systems, but rather the one for nonisolated systems. In this case, let us recall, the entropy of the system can decrease on condition that the entropy of the surroundings increases. Accordingly, there is no paradox, and biological behavior is compatible with the second law.

Now then, the fact that the order observed in living beings can, in principle, be compatible with the second law is by no means an explanation of their structurization. This structurization is allowed, but not explained. Here we come to the second conceptual step: Living systems are not in equilibrium, but are nonequilibrium systems. A system in equilibrium is "dead." For this reason, it is more important to consider the aspects of nonequilibrium in the thermodynamic study of living beings. Furthermore, when the systems move sufficiently far away from equilibrium, structurization appears! Therefore, it is even more attractive to study nonequilibrium phenomena; we can offer an explanation, still very embryonic, for the structurization of living systems.

This is not to say that traditional equilibrium thermodynamics is of no interest to biology. On the contrary: the direction of evolution of processes and the final conditions that the equilibrium has if the system moves toward it are fixed, and consequently it is an indispensable reference point. Nevertheless, since this subject is already a classical subject treated abundantly in the bibliography, we will not pursue it here. In the following, we focus on the study of nonequilibrium processes. In the next chapter, we undertake the study of open systems, that is, those that can exchange matter with the exterior in addition to exchanging energy. Chapter 3 (the last part of the introduction to the theory) is dedicated to nonequilibrium systems. We thus establish the foundations (open and nonequilibrium systems) for the study of processes in biological systems.

BIBLIOGRAPHY

There are many classical thermodynamics texts available to the interested reader. We can cite the following books as a guide, without slighting other valuable and suitable works.

Adkins, C. J., *Equilibrium Thermodynamics*. Maidenhead: McGraw-Hill, 1968.

Callen, H. B., *Thermodynamics and an Introduction to Thermostatistics*, 2nd ed. New York: Wiley, 1985.

MOROWITZ, H. J., *Entropy for Biologists: An Introduction to Thermodynamics*. New York: Academic, 1970.

Other books for a popular audience that are also of great conceptual interest are

PRIGOGINE, I., *¿Solo Una Ilusión?* [*Only an Illusion?*]. Barcelona: Tusquets, 1983.

SCHRÖDINGER, E., *What Is Life?* New York: Macmillan, 1946.

SCHRÖDINGER, E., *¿Qué es la vida? Ment i Matèria* [*What Is Life? Mind and Matter*]. Barcelona: Edicions 62, 1984.

2

Thermodynamic Potentials. Open Systems

2.1 THERMODYNAMIC POTENTIALS

In the last chapter, we have seen that the criteria for deciding whether a process is spontaneous in a thermodynamic system subject to various conditions are the following:

$$(\Delta S)_{U,V} \geq 0$$

$$(\Delta U)_{S,V} \leq 0$$

$$(\Delta H)_{S,p} \leq 0$$

$$(\Delta F)_{T,V} \leq 0$$

$$(\Delta G)_{T,p} \leq 0$$

The functions $S(U, V)$, $U(S, V)$, $H(S, p)$, $F(T, V)$, and $G(T, p)$, which have already been defined in the last chapter and which allow us to express these criteria, are called the *thermodynamic potentials*. Each one contains all the thermodynamic information on the system. This thermodynamic information consists of: (a) specification of the equilibrium state at which the system arrives at the end of a process; (b) specification of the stability of the equilibrium states and prediction of the phase changes; (c) relations between various thermodynamic quantities. As the reader can appreciate, it is very attractive to be able to include all the information about the system in a single function: equations of state, specific heats, compressibilities, phase changes, and so on. In order to be able to appreciate this and also because we will need it later, let us consider some of these questions in some detail.

2.2 GIBBS EQUATION FOR A CLOSED SYSTEM

We have said that the equation $S(U,V)$ contains all the thermodynamic information about the system. The first question raised is how to define this function or, in other words, how to calculate the change in entropy S of a system if we know the change in U and in V in a process.

According to the first law expressed in differential form, that is, for small variations in U, V, and S, we have

$$dU = dq + dw \tag{2.1}$$

In addition, according to expression (1.1) for the work, $dw = -pdV$, and according to the definition of entropy (1.6), $dS = dq/T$, and we obtain

$$dS = dq/T = (dU - dw)/T = T^{-1}dU + pT^{-1}dV \tag{2.2}$$

This relation is called the *Gibbs relation* for a closed system. Later in this chapter we will generalize it to the case of open systems. Next we will look at some consequences of this relation.

(a) Conditions for Thermal Equilibrium

Let us assume a globally isolated system consisting of two parts (subsystems) (Fig. 2.1) that are initially at different temperatures. What will be the final equilibrium state at which the global system arrives if the two subsystems are placed in thermal contact?

Let us start from the idea that the system tends toward the state of maximum entropy. The first mathematical condition for a differentiable function to have a maximum is that its first derivative be zero. In our case, the function to be maximized is the total entropy of the system, which is the sum of the entropies of the two subsystems, since it is an additive function:

$$S_{\text{total}} = S_1(U_1, V_1) + S_2(U_2, V_2) \tag{2.3}$$

In the case we are considering, V_1 and V_2 are constant; on the other

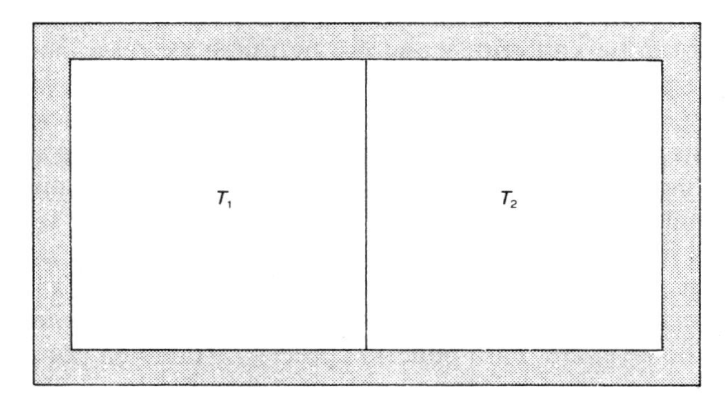

Figure 2.1 Isolated system divided into two subsystems in mutual thermal contact.

hand, the energy of each subsystem varies, with the condition that the restriction

$$U_1 + U_2 = U_{\text{total}} = \text{constant}$$

is satisfied, since the total system is isolated and therefore its energy cannot change. Thus

$$dU_1 + dU_2 = 0$$

since the differential of a constant is zero.

What will be the distribution of the energies, that is, the value of U_1, which has maximum total entropy? According to the maximum condition, differentiating S_{total}, we have

$$dS_{\text{total}} = (\partial S_1/\partial U_1)dU_1 + (\partial S_2/\partial U_2)dU_2$$

and since $dU_1 = -dU_2$

$$dS_{\text{total}} = [(\partial S_1/\partial U_1) - (\partial S_2/\partial U_2)]dU_1 = 0 \text{ (in equilibrium)}$$

In order for the previous relation to be satisfied for any variation in U_1, the difference between the derivatives must be equal to zero. Therefore, in equilibrium we have

$$(\partial S_1/\partial U_1) = (\partial S_2/\partial U_2) \tag{2.4}$$

According to the Gibbs relation (2.2), the derivative of S with respect to U is the reciprocal of the absolute temperature, so that the equality in (2.4) is equivalent to

$$T_1^{-1} = T_2^{-1}$$

That is, the condition for thermal equilibrium is the equality of the temperatures: the two systems exchange energy until they attain the same temperature.

(b) Stability Conditions

The fact that the entropy should be a maximum in an equilibrium state implies first of all that its first derivative must be zero—that is, that the first differential dS must be zero for any value of the variations in the energy of the subsystems. This is a necessary but not a sufficient condition for characterization of a maximum. In addition, it is necessary that the second derivative be negative, that is,

$$d^2S \leqslant 0 \qquad \text{(stability condition)} \tag{2.5}$$

What implications does this stability condition entail in our previous example of thermal equilibrium? If we calculate the second derivative of

S with respect to U, we have

$$(\partial^2 S/\partial U^2) = (\partial T^{-1}/\partial U)_V = -T^{-2}(\partial T/\partial U)_V$$

since the first derivative is

$$(\partial S/\partial U) = T^{-1}$$

according to the Gibbs relation. The stability condition then implies that

$$-T^{-2}(\partial T/\partial U)_V \leqslant 0 \tag{2.6}$$

The derivative $(\partial U/\partial T)_V$ is precisely the heat capacity at constant volume for the system. Thus, the stability condition demands that the heat capacity (and therefore the specific heat) be positive, a result that has been well confirmed by experiment.

In Chapter 9 we again examine in more depth the theory of stability for nonequilibrium stationary states, and we study the significance of instabilities like phase changes.

(c) Equations of State

In the first chapter, we presented the thermometric equation of state (which relates the pressure p to the volume V and the temperature T) and the calorimetric equation of state (which relates the internal energy U to V and T). As we said then, these equations describe the thermodynamic behavior of the system (specific heat, compressibility, phase transition temperatures, . . .). These equations are contained in the fundamental equation $S(U, V)$. In fact, the derivatives of S with respect to these variables are

$$\begin{aligned} T^{-1} &= (\partial S/\partial U)_V = T^{-1}(U, V) \\ pT^{-1} &= (\partial S/\partial V)_U = pT^{-1}(U, V) \end{aligned} \tag{2.7}$$

These equations are not yet the thermometric and calorimetric equations of state, but we can immediately obtain them. If we rearrange the first equation (2.7), we obtain a function $U = U(T, V)$, which is the calorimetric equation of state. If we next introduce this equation into the second equation, we arrive at a relation among p, T, and V which is the thermometric equation of state $p = p(T, V)$. This demonstrates that all the experimental information about the systems included in the thermometric and calorimetric equations of state is contained in the fundamental equation or the thermodynamic potential $S(U, V)$. In addition, the equality of the cross mixed derivatives

$$(\partial^2 S/\partial V \partial U) = (\partial^2 S/\partial U \partial V)$$

established by means of purely mathematical considerations, leads to a series of relations between the derivatives of the equations of state which are of great experimental interest. Starting from these relations and knowing certain functions, we can obtain other functions by purely mathematical procedures without needing to make a major experimental effort.

(d) Differential Relations for Other Potentials

Starting from the Gibbs relation (2.2) and the definition of the other thermodynamic potentials, we can immediately obtain their respective differential relations. In the case of the internal energy $U = U(S, V)$, and rearranging relation (2.2), we obtain

$$dU = TdS - pdV \tag{2.8}$$

For the enthalpy $H = U + pV$, we have $dH = dU + pdV + Vdp$. Introducing (2.8) into this relation, we obtain

$$dH = TdS + Vdp \tag{2.9}$$

For the Helmholtz function $F = U - TS$, we find

$$dF = dU - TdS - SdT$$

and considering (2.8), we immediately arrive at

$$dF = -SdT - pdV \tag{2.10}$$

Finally, for the free energy G or the Gibbs function defined as

$$G = U + pV - TS$$

we have

$$dG = dU + pdV + Vdp - TdS - SdT$$

which combined with (2.8) leads to

$$dG = -SdT + Vdp \tag{2.11}$$

Any of these potentials, expressed in terms of their respective variables, contains all the thermodynamic information about the system, just like the function $S(U, V)$. We can go from one potential to another by means of what is called a Legendre transformation, which we are not interested in considering in detail. Let us finally observe that these relations are valid only for closed systems. Biological systems are open systems; that is, they can exchange matter in addition to energy. For this reason, we need to study how these relations are modified in systems with these characteristics.

2.3 GIBBS RELATION IN OPEN SYSTEMS

Open systems are characterized by exchange of matter, either with the exterior through the walls or within the system proper when, for example, one substance disappears and others appear due to chemical reactions. In this case, it is not sufficient to use the variables U and V; rather, information about the composition of the system must be included. Usually this is accomplished by indicating N_i, that is, the number of moles of each chemical species i present in the system.

The thermodynamic potentials become a function of, for example, N_i, $S(U, V, N_i)$, and $U(S, V, N_i)$. In order to describe the effects of variations in the composition of the system, we define the *chemical potential* of the species i as

$$\mu_i = (\partial U/\partial N_i)_{S,V,N_{j\neq i}} \tag{2.12}$$

where the subscripts S, V, and N_j indicate that the entropy, the volume, and the number of moles of the rest of the substances are held constant while U is varied. Despite the fact that it is a less familiar variable, the chemical potential plays a role analogous to pressure and temperature, which are much more commonly used and are more intuitive. Thus, just as heat passes from zones of higher temperature to zones of lower temperature until the temperatures equalize, we will see that substances pass from zones in which the chemical potential is high to those in which it is low. This is the origin of the name "chemical potential": it is an analogy to the gravitational potential, in terms of which masses tend to go from zones with a high potential (high height) to zones with a low potential (low height).

According to definition (2.12), the differential of the internal energy is written

$$dU = TdS - pdV + \Sigma_i \mu_i dN_i \tag{2.13}$$

where Σ_i indicates summation of the product μdN for all the i chemical species that compose the system. This relation is logical, since now we cannot vary only S and V [as in the case of a closed system in which the equation is limited to (2.8)], but in addition we must consider the possible variations in N_i, which must be added to the corresponding variations in S and V.

Starting from (2.13), we determine the differential relation corresponding to the entropy $S(U, V, N_i)$ by solving for dS.

$$dS = T^{-1}dU + pT^{-1}dV - \Sigma_i \mu_i T^{-1} dN_i \tag{2.14}$$

The relations (2.13) and (2.14), which generalize (2.8) and (2.2) respectively, are the Gibbs equation proper. It was Gibbs, an American scientist, who in 1876 introduced the concept of chemical potential and made the greatest possible use of the information contained in these differential relations.

2.4 THE CHEMICAL POTENTIAL

The concept of chemical potential appears extensively throughout this text. Therefore it is expedient to familiarize ourselves as much as possible with its use. Just as there are thermometric and calorimetric equations of state, there are also equations of state for the relations between the chemical potentials and T, V, and N_i.

$$\mu_i = \mu_i(T, V, N_i)$$

The system most commonly used at the elementary level is the ideal gas, whose thermometric and calorimetric equations of state are respectively, in the monatomic case,

$$p = NRT/V \tag{2.15}$$

$$U = (3/2)NRT \tag{2.16}$$

On the other hand, the equation of state for the chemical potential in ideal gases,

$$\mu_i(T, V, N_i) = \mu_i^0(T) + RT \ln p_i \tag{2.17}$$

is a much less familiar result. In the last expression, p_i is the partial pressure, that is, $p_i = N_iRT/V$, and μ_i^0 is a function that depends only on the temperature and not on the concentration.

Expression (2.17) corresponds to the chemical potential of an ideal gas. If we recall that in ideal gases the partial pressure is the total pressure p multiplied by x_i, the mole fraction, defined as

$$x_i = N_i/N$$

the quotient of the number of moles of species i divided by the total number of moles $N = \Sigma_i N_i$ of all the chemical species of the system, the chemical potential can be expressed as

$$\begin{aligned} \mu_i(T, p, N_i) &= \mu_i^0(T) + RT \ln px_i \\ &= \mu_i^0(T + RT \ln p + RT \ln x_i \end{aligned} \tag{2.18}$$

If we include the term $\mu_i^0 + RT \ln p$ in a single function, we have

$$\mu_i(T, p, N_i) = \mu_i^*(T,p) + RT \ln x_i \tag{2.19}$$

There is another possibility: We can describe it as a function of the concentration of each species $c_i = N_i/V$:

$$\mu_i(T, V, c_i) = \mu_i(T, V) + RT \ln c_i \tag{2.20}$$

We discuss such a diversity of equivalent formulations because in different parts of the text, we use different formulations, depending upon individual cases.

In the case of nonideal systems, the equation of state for the chemical potential (2.19) is transformed to

$$\mu(T, p, N_i) = \mu_i^*(T, p) + RT \ln \gamma_i x_i \tag{2.21}$$

where γ_i is the activity coefficient, which is generally a complicated function of T, p, and N_i. For an ideal system, however,

$$\gamma_i = 1$$

Let us explain that the relations (2.19), (2.20), and (2.21) are applicable not only to mixtures of gases, but also to liquid solutions, which is the case we study in greater detail. On the other hand, (2.17) and (2.18) are applicable only to mixtures of ideal gases.

Finally, in the presence of external fields, the potential energy per mole (or per unit mass) due to the external field is included in the chemical potential. This fact accentuates even more the previously mentioned relationship between the chemical potential and the gravitational or electric potential. In a gravitational field, the chemical potential is expressed as

$$\mu_i(T, p, N_i) = \mu_i^*(T, p) + RT \ln \gamma_i x_i + M_i gh \tag{2.22}$$

where M_i is the molar mass.

When the external field is an electric field, we have

$$\mu_i(T, p, N_i) = \mu_i^*(T, {}_p) + RT \ln \gamma_i x_i + \mathcal{F} z_i \phi \tag{2.23}$$

where $\mathcal{F}$ is the faraday, or the electric charge per mole ($\mathcal{F} = 96{,}500$ coulombs/mole), z_i is the valence (number of missing or extra electrons) of the chemical species i, and ϕ is the electric potential. The term $\mathcal{F} z_i \phi$ then represents the potential energy per mole.

Finally, a centrifugal field, that is, when the material is in a centrifuge with rotational velocity ω, its chemical potential is written as

$$\mu_i(T, p, N_i) = \mu_i^*(T, p) + RT \ln \gamma_i x_i - M_i(1 - \rho v_i)\omega^2 r^2/2) \tag{2.24}$$

In the last expression, M_i is the molar mass of species i, r is the position or radius of gyration of the particles, ρ is the density of the solvent, and v_i is the specific partial volume of species i (reciprocal of the density).

For the purpose of this book, it is not essential to justify these expressions, just as we have not justified the thermometric and calorimetric equations of the ideal gas. In subsequent chapters, we recall these relations in the most convenient form for our study.

2.5 PHASE EQUILIBRIUM

In the last section, we indicated that the chemical potential plays a role analogous to the temperature: Just as heat passes from an elevated temperature T to a low temperature until the temperatures of the system equalize, matter passes from zones of high chemical potential to zones of low chemical potential until the chemical potentials equalize. Let us prove this statement.

Since in this text we are very often involved with conditions of constant T and p (the most common conditions in the laboratory), let us go to a thermodynamic potential that is more appropriate for working under these circumstances—that is, the free energy ($G(T, p, N_i)$. We know that for a process to be possible, it needs to allow for a decrease in G. Consequently, in the equilibrium state (that is, the state in which the system no longer experiences further evolution unless the conditions change), G should reach a minimum and (just as in isolated systems) the system tends toward the state with maximum entropy. The necessary condition for the minimum, like that for the maximum, is that the first derivative (or the combination of first derivatives) be zero. The reasoning is analogous to that used in the second section to study thermal equilibium in an isolated system. In our case, dG can be given by

$$dG = -SdT + Vdp + \Sigma_i \mu_i dN_i \tag{2.25}$$

This relation, which generalizes (2.11), is obtained by starting from the Gibbs equation for U (2.13) and the definition of G,

$$G = U + pV - TS$$

according to which

$$dG = dU - TdS - SdT + pdV + Vdp$$

Since now T and p are constant, $dT = dp = 0$, and (2.25) for each subsystem is reduced to

$$dG = \Sigma_i \mu_i dN_i$$

The total Gibbs function for a system composed of two subsystems α and β in material contact, that is, with the possibility of matter exchange (Fig. 2.2), is

$$G_{\text{total}} = G_\alpha + G_\beta$$

and the necessary condition for a process to be able to occur in a system is

$$dG_{\text{total}} = dG_\alpha + dG_\beta \leqslant 0 \tag{2.26}$$

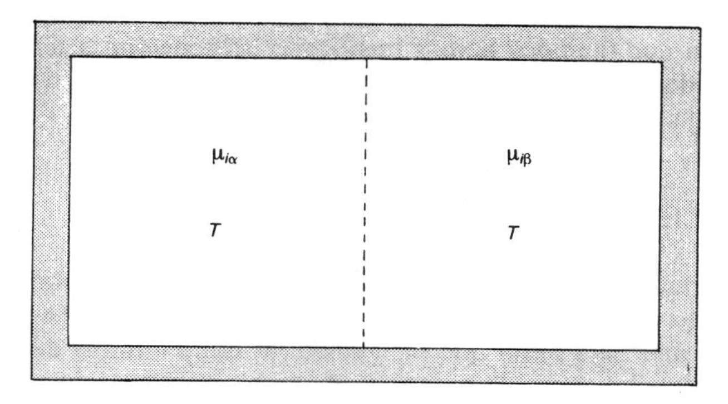

Figure 2.2 Isolated system divided into two subsystems in thermal and material contact (they can exchange heat and matter).

for which we have

$$dG_{\text{total}} = \Sigma_i \mu_{i\alpha} dN_{i\alpha} + \Sigma_i \mu_{i\beta} dN_{i\beta} \leqslant 0 \tag{2.27}$$

If the global system is closed (that is, the substances can pass from α to β and vice versa, but cannot leave the combined container), for each chemical species the total number of moles in α plus the total number of moles in β must remain constant (we assume for now that there are no chemical reactions). Therefore,

$$dN_{i\alpha} + dN_{i\beta} = 0$$

and then

$$\Sigma_i (\mu_{i\alpha} - \mu_{i\beta}) dN_{i\alpha} \leqslant 0 \tag{2.28}$$

The equilibrium condition is reduced to the last expression being an equality. Since the variations in different species are independent, as an equilibrium condition we have

$$\mu_{i\alpha} = \mu_i \beta \tag{2.29}$$

Thus, the equilibrium condition for material transport is analogous to the condition for thermal equilibrium. In the latter case, the condition is equality of the temperatures. In the former, the condition is equality of the chemical potentials for each species on both sides of the wall. This does not mean that the chemical potential of all the species must be the same, but that the chemical potential of each species individually in compartment α must be equal to the chemical potential of the same species in compartment β.

Finally, if $\mu_{i\alpha} \geqslant \mu_{i\beta}$, for a process to be able to occur it is necessary that $dN_{i\alpha} \leqslant 0$; that is, the quantity of species i in α must decrease and the quanitity of species i in β must increase. In other words, if $\mu_{i\alpha} > \mu_{i\beta}$, there must be a flow or transport of species i from α toward β; that is, matter circulates from zones of elevated chemical potential to those of low chemical potential. We have thus proven our previous statement, and we may consider the intuitive idea of chemical potential which we presented at the beginning to be valid.

Condition (2.29) is extremely useful. Starting from this condition, we can obtain expressions for the osmotic pressure, freezing point depression, boiling point elevation, vapor pressure, triple point conditions, and many other results which we will not repeat since they are familiar from elementary chemistry or physical chemistry.

2.6 CHEMICAL EQUILIBRIUM

We conclude this chapter by considering the case of a system where chemical reactions occur. In this case, the matter, instead of passing from one side to the other (and changing the composition of the system by entering or leaving), undergoes chemical transformations. The basic problem to resolve in this system is: What will be the composition of the system when the chemical reaction reaches equilibrium, and how do we change these conditions in order to vary (for example) the temperature T or the pressure p? The answer can be given for the same criterion as before: minimization of the free energy G and constant T and p. As we have seen, this implies that the first differential of G is zero; with this, starting from (2.25) and with $dT = dp = 0$, we obtain

$$dG = \Sigma_i \mu_i dN_i = 0 \tag{2.30}$$

Now we do not consider the system to be divided into parts. This is not necessary, since we assume that the different values of N_i vary due to the chemical reactions.

In this case, however, the variations dN_i are not independent, but rather are related by the stoichiometry of the specific chemical reaction. Let us assume the following chemical reaction:

$$\nu_A A + \nu_B B + \cdots \longrightarrow \nu_C C + \nu_D D + \cdots \tag{2.31}$$

This means that for each ν_A moles of A and for each ν_B moles of B which disappear in the reaction, ν_C moles of C and ν_D moles of D appear. This relation can be written in mathematical form as

$$\frac{dN_A}{\nu_A} = \frac{dN_B}{\nu_B} = \frac{dN_C}{\nu_C} = \frac{dN_D}{\nu_D} = \cdots = d\xi \tag{2.32}$$

the parameter ξ is called *the extent of reaction,* and tells us how many moles of each species has reacted. In other words, due to the fact that the variations dN_i of the different species are not independent, we have been able to simplify the problem by reducing the different dN_i to a single quantity $d\xi$. In the expression (2.32), we have considered that the stoichiometric coefficients of the products ($C, D, \ldots$) are positive, since substances C and D appear when the reaction proceeds in the forward direction (to the right), and the stoichiometric coefficients of the reagents ($A, B, \ldots$) are negative, since they disappear when the reaction occurs in the indicated direction.

Given that according to (2.32), $dN_i = \nu_i d\xi$, we can write (2.30) in the form

$$\Sigma_i \mu_i dN_i = \Sigma_i \mu_i \nu_i d\xi = 0 \tag{2.33}$$

and since $d\xi$ can attain any value, the condition for chemical equilibrium is

$$\Sigma_i \nu_i \mu_i = 0 \tag{2.34}$$

For ideal gases, in which (2.17) is satisfied for the chemical potential, this condition leads to

$$\Sigma_i \mu_i^0(T)\nu_i + RT\Sigma_i \nu_i \ln p_i = 0$$

that is,

$$RT \ln \Pi_i p_i^{\nu_i} = -\Sigma_i \mu_i^0(T)\nu_i$$

or else, finally,

$$\Pi_i p_i^{\nu_i} = K(T) \tag{2.35}$$

where the symbol Π is used to indicate the product of the values of the i terms which follow it.

This is the usual condition for chemical equilibrium in reactions between gases, well known in elementary chemistry, where $K(T)$ is the equilibrium constant. We observe that $K(T)$ depends on the temperature, and in this sense it is not strictly a constant. The name *equilibrium constant* refers to the fact that it does not depend on the concentrations, which are the unknowns that concern us. According to (2.35), the equilibrium constant $K(T)$ can be given by

$$K(T) = \exp\left[-(1/RT)\Sigma_i \nu_i \mu_i^0(T)\right] \tag{2.36}$$

This relation makes it obvious why we are interested in the functions μ_i^0, which determine the equilibrium constant.

In elementary chemistry or physical chemistry texts, other possible forms are given for the chemical equilibrium condition as a function of the mole fractions, the concentrations, and so on, and the variation of $K(T)$ with temperature is studied. We will not repeat these studies here, but we emphasize that the different forms for writing the equilibrium condition are exactly equivalent to the general relation (2.34), which can lead to very complicated expressions in the case of nonideal systems.

In the chapters dealing with applications to biological systems, we will pay special (almost exclusive) attention to phenomena of material transport and chemical reactions, in which the chemical potential plays a predominant role. For this reason, we have devoted this chapter to a review of the chemical potential and the conditions for material transport equilibrium and chemical equilibrium.

BIBLIOGRAPHY

Here we cite classical texts which stress the thermodynamic description of phenomena associated with chemical transformations.

CALLEN, H. B., *Thermodynamics and an Introduction to Thermostatistics*, 2nd ed. New York: Wiley, 1985.

CRIADO-SANCHO, *Introducción Conceptual a la Termodinámica Química* [*Conceptual Introduction to Chemical Thermodynamics*]. Madrid: AC, 1983.

GLASSTONE, S., *Thermodynamics for Chemists*. New York: D. Van Nostrand, 1946.

GRAY, H. B., and G. P. HAIGHT, Jr., *Basic Principles of Chemistry*. New York: Benjamin, 1967.

TEJERINA, F., *Termodinámica* [*Thermodynamics*]. Madrid: Paraninfo, 1976.

3

Thermodynamics of Irreversible Processes

3.1 INTRODUCTION

Up to now, we have studied the criterion which tells us whether or not, given any two equilibrium states, passing from the first to the second is possible under the conditions to which the system is subjected. On the other hand, we have not studied the question concerning the rate at which the possible processes between the states evolve. Obviously, the latter information is also of great importance. Let us present a practical example. In order for an exothermic chemical reaction (that is, one which releases energy) to be carried out to the greatest degree of completion (in order to obtain the maximum quantity of products from a certain quantity of reactants), it is advisable to perform the reaction at low temperatures. This is the information that is provided by equilibrium thermodynamics. However, the reaction is very slow at low temperatures. For practical purposes (for example, industrial), we need to arrive at a compromise between these two factors: Maximum thermodynamic efficiency advises the lowest possible temperature, while kinetic considerations advise against lowering the temperature. Therefore, we cannot neglect either of the two pieces of information, but equilibrium thermodynamics provides only the first piece.

It is then essential to go on to studying the kinetic factors. This is one of the objectives of the thermodynamics of irreversible processes. Another objective is to study the nonequilibrium stationary states. We have already mentioned that living systems are found in nonequilibrium states. Exchange of ions across the cell membrane is a good example of this. Sodium tends to enter the cell both by diffusion (its concentration is greater outside than inside the cell) and by electrical effects (the electrical potential inside is less than outside). In order to maintain this stationary state, the cell membrane has available a molecular pump, which allows it to export all the sodium that enters. If any inhibitor stops the functioning of this pump, the cell fills up with sodium until it reaches an equilibrium situation, which often represents cellular death. Therefore, even in such a simple aspect of cellular activity, the cell at rest is not found in an equilibrium state but rather in a nonequilibrium stationary state. The difference between an equilibrium state and a nonequilibrium stationary state is that in order to maintain the latter, energy must be continuously supplied to the system, which is not necessary when the state is an equilibrium state. The energy necessary to maintain the cells in their nonequilibrium states is provided by the cell metabolism. Therefore, it is essential to take into consideration the nonequilibrium stationary states in order to determine the amount of energy per unit time needed by a cell and to determine on which factors this energy depends.

3.2 DISCONTINUOUS SYSTEMS. PRODUCTION OF ENTROPY

The object of this chapter is to study the general ideas of the linear thermodynamics of irreversible processes. In general, we know that the variation in entropy dS is defined by

$$dS = dq_r/T$$

where dq_r is the heat exchanged between the system and the medium when the system evolves ideally and reversibly from the initial state to the final state. If, instead of using the heat dq_r exchanged in the ideal reversible process, we use the heat dq exchanged in the real process, we have

$$dS \geq dq/T = d_eS \tag{3.1}$$

The equality applies only if the process is reversible. We have denoted as d_eS the exchange of entropy of the system with the exterior. The inequality (3.1) can be written in the form of an equality as

$$dS = d_iS + d_eS \tag{3.2}$$

where d_iS represents the entropy produced in the system proper. This equality expresses the fact that the entropy of the system can change for two reasons: (1) By transport of entropy across the walls of the system (d_eS) (this exchange can be positive if the entropy enters from the exterior, or negative if the system transfers entropy to the exterior). (2) There is a production or creation of entropy in the interior of the system (d_iS) (this term, according to the second law of thermodynamics, must be positive). In a reversible process, $d_iS = 0$; that is, the variation in the entropy should be exclusively due to exchange with the exterior, without the total entropy's increasing. In an irreversible process, on the other hand, $d_iS \geq 0$. It is observed that in an isolated system, in which exchange with the exterior is impossible, $d_eS = 0$. Then

$$dS = d_iS \geq 0$$

and we recover the criterion according to which the entropy can increase only in an isolated system.

Now let us assume, as in the examples studied in the last chapter, the case of an isolated system divided into two compartments of fixed volume which can exchange energy and matter (Fig. 3.1). In this case, we can express the change in the total internal energy as a function of the changes in the internal energy in each subsystem. Since the global system is isolated, the total internal energy is constant and we have

$$dU = dU_1 + dU_2 = 0$$

If we also assume that the total number of moles of each compartment is constant (since there are no chemical reactions in the system, but only transport), we have

$$dN_i = dN_{i1} + dN_{i2} = 0$$

for all the components i of the system. We want to study the question of which processes are possible in the interior of this isolated system. The criterion can be given for the variation in the total entropy of the system

$$dS = dS_1 + dS_2 \geq 0$$

According to the Gibbs equation (2.14)

$$dS = T^{-1}dU + pT^{-1}dV - \Sigma_i \mu_i T^{-1} dN_i$$

where T is the absolute temperature and μ_i is the chemical potential of species i. Just as in the examples cited in the last chapter, we now have

$$dS = (T_1^{-1} - T_2^{-1})dU_1 - \Sigma_i(\mu_{i1}T_1^{-1} - \mu_{i2}T_2^{-1})dN_{i1} \geq 0 \qquad (3.3)$$

which can be written in the simpler form

$$dS = \Delta(T^{-1})dU_1 - \Sigma\Delta(\mu_i T^{-1})dN_{i1} \geq 0 \qquad (3.4)$$

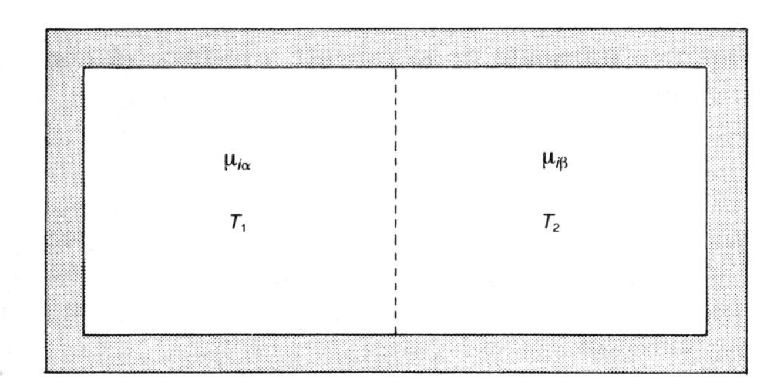

Figure 3.1 Isolated system divided into two subsystems in thermal and material contact (they can exchange heat and matter).

where

$$\Delta(T^{-1}) \equiv (T_1^{-1} - T_2^{-1})$$

$$\Delta(\mu_i T^{-1}) \equiv (\mu_{i1} T_1^{-1} - \mu_{i2} T_2^{-1})$$

Condition (3.4) expresses in mathematical form two well-known facts: (1) Heat is transmitted from the hot to the cold since if $T_1 > T_2$ then

$$\Delta T^{-1} < 0$$

and therefore $dU_1 < 0$; that is, the hotter system loses energy in the form of heat, which is transferred to the colder system. (2) Matter is transferred from the zones of higher chemical potential to those of lower chemical potential, since if $\mu_{i1} > \mu_{i2}$, then

$$\Delta(\mu_i T^{-1}) > 0$$

and $dN_{i1} < 0$, which means that subsystem (1) loses matter of species i, which is transferred to subsystem (2).

The moment has arrived for time to appear in this text. We observe that time is not involved in equilibrium thermodynamics, an infrequent circumstance in physics. As we have said, the thermodynamics of irreversible processes is concerned with the rate of processes and must involve time. Since entropy plays a central role in thermodynamics, we will first ask how entropy varies with time.

Differentiating (3.4) with respect to time, we obtain

$$dS/dt = \Delta(T^{-1})dU_1/dt - \Sigma_i \Delta(\mu_i T^{-1})(dN_{i1}/dt) \geq 0 \tag{3.5}$$

We observe that this expression is a sum of products; that is, it is a bilinear form. We define the *thermodynamic fluxes* of energy and matter in terms of their respective time derivatives

$$\begin{aligned} J_u &= A^{-1}(dU/dt) \\ J_i &= A^{-1}(dN_i/dt) \end{aligned} \tag{3.6}$$

where A is the area of the cross section normal to the flow. A flux then indicates the variation of a quantity per unit area and time. In other words, it indicates how much energy or matter crosses the boundary between the two subsystems (Fig. 3.1) per unit area and per unit time.

We call the imbalances that cause the fluxes the *conjugate thermodynamic forces*. In the example that we will consider, the forces conjugate to the dissipative fluxes are

$$\begin{aligned} X_u &= \Delta(T^{-1}) \\ X_i &= -\Delta(\mu_i T^{-1}) \end{aligned} \tag{3.7}$$

As we have said, the forces indicate imbalance (that is, temperature difference, difference in chemical potential, and so on) between the two compartments of the system. As a response to these imbalances, the dissipative fluxes J_u, J_i, and so forth arise.

The variation in the total entropy of the system or the *production of entropy* (entropy that is created per unit time, in this case), given by (3.5), can be written now as

$$dS/dt = A(J_u X_u + \Sigma_i J_i X_i) \geq 0 \tag{3.8}$$

3.3 CONSTITUTIVE EQUATIONS

We call the relations between the thermodynamic fluxes and forces the *constitutive equations*. These equations are central to the thermodynamics of irreversible processes, since relating the fluxes (time derivatives) to the forces (degree of imbalance) gives the rate of the processes as a function of the degree of imbalance. The constitutive equations depend on the system being studied, that is, on the considered material and occasionally on its temperature and composition. Thermodynamics, as a purely macroscopic science, cannot indicate the exact form of these equations but supplies very valuable information [by means of the second law expressed in form (3.8)] which allows us to discard equations that do not satisfy the restrictions of the second law. Of course, this is a great help, since it allows us to avoid wasting time considering possible equations that are not plausible physically.

In classical thermodynamics of irreversible processes, the constitutive equations are considered to be linear. Therefore we have

$$J_u = A_{uu} X_u + \Sigma_i A_{ui} X_i \tag{3.9}$$

$$J_i = A_{iu} X_u + \Sigma_j A_{ij} X_j \tag{3.10}$$

The coefficients A_{uu}, A_{ui}, A_{iu}, and A_{ij} are called *phenomenological coefficients*; they can be constant or functions of the temperature and composition of the system, but not functions of the degree of imbalance. In the final part of this book we will study some consequences that arise when nonlinear terms appear in the constitutive equation, that is, when terms of the type $X_u X_i$ or $X_i X_j$ are involved.

As a simple example of linear constitutive equations, we can consider the case of a system in which there is only thermal imbalance. Then Eq. (3.9) is reduced to

$$J_u = A_{uu} \Delta(T^{-1})$$

which, with the definition of $\Delta(T^{-1})$, leads to

$$J_u = A_{uu}(T_1^{-1} - T_2^{-1})$$

which also can be written as

$$J_u = K(T_2 - T_1) \tag{3.11}$$

if we define $K \equiv A_{uu}/T_1T_2$. This law is called *Newton's law of cooling*, and establishes that the heat flux is proportional to the temperature difference.

According to the definition (3.6) of J_u, (3.11) can be written in the form

$$dU_1/dt = AK(T_2 - T_1)$$

and since $dU_1 = m_1c_1dT_1$, by the definition of specific heat c_1 per unit mass for system (1), we have

$$dU_1/dt = m_1c_1(dT_1/dt) = AK(T_2 - T_1) \tag{3.12}$$

If, for example, we assume that T_2 is constant [let us imagine that subsystem (1) is a hot container initially at temperature $T_1(0)$ while subsystem (2) is the laboratory at constant temperature T_2], we can determine the temperature variation of subsystem (1) by integrating the last equation and obtaining

$$T_1(t) = T_2 - (T_2 - T_1(0)) \exp[-(AK/m_1c_1)t] \tag{3.13}$$

This relation describes the decrease in temperature of a hot object that has been allowed to cool down in a room. As we see, the temperature decreases exponentially until it is equal to the ambient temperature. If we do two different experiments, we can determine the specific heat c_1 and the constant K, since A (in this case, the area of the container) and m_1 (its mass) can be measured more or less easily.

3.4 SECOND-LAW RESTRICTIONS ON THE CONSTITUTIVE EQUATIONS

If we now introduce the constitutive equations (3.9) and (3.10) into the entropy production (3.8), the positive nature of the latter imposes a series of restrictions on the phenomenological coefficients. In the present example, restricted to heat conduction, we have

$$dS/dt = AJ_uX_u = AA_{uu}X_u^2 \geq 0$$

Since A and X_u^2 are always positive, the coefficient A_{uu} will also be positive. This agrees with experiment, since it implies that the constant K in (3.11) is positive and therefore that heat is transmitted from the hot to the cold.

Introduction of the constitutive equations in the case when there is only energy transport and a single chemical component leads to

$$dS/dT = A[A_{uu}X_u^2 + A_{ii}X_i^2 + (A_{ui} + A_{iu})X_uX_i] \geq 0$$

In order for this expression to be positive for each value of X_u and X_i, the following restrictions must be satisfied:

$$
\begin{aligned}
A_{uu} &\geq 0 \\
A_{ii} &\geq 0 \\
A_{uu}A_{ii} &\geq A_{ui}A_{iu}
\end{aligned}
\tag{3.14}
$$

The first of these conditions has already been treated in this section. The second is immediately obtained if we assume that $X_u = 0$: in this case, the positive nature of X_i^2 implies that the inequality can be satisfied only if $A_{ii} \geq 0$. The third condition expresses the fact that the determinant of the bilinear expression must be positive. This is one of the conditions that must be satisfied by a matrix in order to be positive definite. The conclusion is that the second law, which suggested the form of (3.9) and (3.10) for the constitutive equations, in addition establishes some restrictions on their coefficients and therefore in practice provides a great deal of guidance concerning whether or not the equations and behaviors will be physically possible.

3.5 ONSAGER RELATIONS

The physicist Lars Onsager proved in the year 1931 that the matrix of the phenomenological coefficients is symmetric. In other words, he established the relation

$$
\begin{aligned}
A_{ui} &= A_{iu} \\
A_{ij} &= A_{ji}
\end{aligned}
\tag{3.15}
$$

for the coefficients (also called dissipative coefficients). This result is very important, since it reduces the number of independent coefficients. Thus if we consider only energy transport for a single chemical species, we have the equations

$$
\begin{aligned}
J_u &= A_{uu}X_u + A_{ui}X_i \\
J_i &= A_{iu}X_u + A_{ii}X_i
\end{aligned}
$$

with four coefficients (A_{uu}, A_{iu}, A_{ui}, A_{ii}). According to Onsager $A_{ui} = A_{iu}$, with which the independent parameters are reduced to three. From an experimental point of view, this is important, since measurement of A_{ui} gives information about A_{iu}, and therefore in principle it is not necessary to carry out a series of experiments in order to measure the latter coefficient. This allows a great savings in time, and makes the work of the investigator much easier.

These relations, for which Onsager received the Nobel Prize in Chemistry in 1976, are one of the most important results in the thermodynamics of irreversible processes. In fact, relations (3.15) are equalities, that is, they give exact information, while purely thermodynamic considerations from the second law can only establish the inequalities in (3.14), which obviously contain less information than an equality. Nevertheless, the *Onsager relations* cannot be deduced by starting from strictly thermodynamic considerations, but rather other considerations are essential: microscopic, statistical mechanics, and fluctuation theory. The Onsager reciprocity relations have been experimentally confirmed many times and summarize results known and used previously in various areas, such as in chemistry (detailed balance) or in thermoelectricity. From a thermodynamic point of view, there are authors who consider the Onsager relations as an authentic "fourth law" of thermodynamics.

It is very important to note that these relations are satisfied only when the fluxes are expressed as a function of their respective conjugate thermodynamic forces. This is the reason why it is so important to know the precise form of the entropy production, since without it we do not know which combination of forces should be assigned to a given combination of fluxes in order for the Onsager relations to be valid.

3.6 CONTINUOUS SYSTEMS

Imagine that, instead of a system divided into two homogeneous compartments, we have a system in which the thermodynamic variables vary continuously. In this case, instead of the temperature difference or chemical potential difference, we take their spatial derivatives to be the thermodynamic forces; that is,

$$\begin{aligned} \Delta(T^{-1}) &= (dT^{-1})/dx)dx \\ \Delta(\mu T^{-1}) &= [d(\mu T^{-1})/dx]dx \end{aligned} \tag{3.16}$$

where we assume that x is the direction in which the values of the indicated thermodynamic quantities vary. The entropy production under these conditions can be given by

$$dS/dt = A[J_u(dT^{-1}/dx) - \Sigma_i d(\mu_i T^{-1})/dx]dx$$

It is useful to define the entropy production per unit volume and time as

$$\sigma = (1/dV)(dS/dt)$$

and since $Adx = dV$, we arrive at

$$\sigma = J_u(dT^{-1}/dx) - \Sigma_i d(\mu_i T^{-1})/dx \tag{3.17}$$

When the thermodynamic properties do not vary in a single direction, we replace d/dx in the previous equations by the gradient operator ∇, which is a vector whose components are the derivatives in the respective directions of the axes,

$$\nabla = (d/dx, d/dy, d/dz)$$

In the following we will refer only to one-dimensional problems, but the generalization of the theory to the three-dimensional case is immediate.

In the continuous case, we then have the following combinations of fluxes and forces:

Forces	**Fluxes**
$X_u = dT^{-1}/dx$	J_u
$X_i = -d(\mu_i T^{-1})/dx$	J_i

(3.18)

The constitutive laws and the Onsager relations hold under the conditions imposed just as we have outlined in the case of discontinuous systems. As an example, let us write the law corresponding to heat transport in the case in which we can neglect material transport.

$$J_u = A_{uu}(dT^{-1}/dx)$$

where A_{uu} is the corresponding phenomenological coefficient, which must be positive according to the restrictions of the second law. Comparing with the *classical Fourier law* (1812) for heat conduction,

$$J_u = -\lambda(dT/dx) \tag{3.19}$$

where λ is the thermal conductivity, we make the identification

$$\lambda = A_{uu}T^{-2}$$

and in this way we relate the phenomenological coefficient to a very familiar variable. We observe in passing that the familiar transport laws for many years have been automatically included within the core of the linear formulation of the thermodynamics of irreversible processes, which in more complicated cases provides convenient generalizations. We also note that the structure of the theory in the continuous case is exactly the same as in the discrete case, with a minimal change in the definition of the thermodynamic forces.

In Table 3.1, we present some values of the thermal conductivity, a property which depends on the material and the temperature. In this table we can see, in more significant detail, the great difference between a good conductor of heat and electricity like copper and various thermal insulators like cork, air, or animal fat, used in architecture or by animals to manage thermal regulation nearly independent of the exterior temperature.

TABLE 3.1 Thermal Conductivity of Different Materials

Material	Values of λ ($W/m{\cdot}K$)
Copper	365.0
Cork	0.05
Water	0.06
Air	0.03
Animal fat	0.01

The evaluation of these coefficients for each material corresponds to nonequilibrium statistical mechanics. Thermodynamics cannot give more information than the fact that these coefficients are positive. The situation is analogous to what occurs in equilibrium with the specific heats: Thermodynamics can only state that they are positive and lead to certain relations between them, but it is unable to give concrete values.

3.7 MINIMUM ENTROPY PRODUCTION

In the chapter devoted to equilibrium thermodynamics, we saw that systems tend to maximize the entropy or to minimize the free energy or other thermodynamic potentials, according to the conditions under which they are found. Prigogine demonstrated about 1945 that, under certain conditions (linear phenomenological laws, constant phenomenological coefficients, Onsager reciprocity relations), entropy production in stationary states is minimum. The importance of this theorem involves the fact that it allows us to consider, in a certain way, the production of entropy as a nonequilibrium thermodynamic potential, both for determining the stationary states and for studying their stability. On the other hand, the numerous restrictions that limit this theorem make it much less general than the classical results of equilibrium theory. At present, no nonequilibrium potential is known that has generality comparable to that of the equilibrium potentials, nor is such a potential in fact believed to exist.

If we represent the entropy production per unit time and volume, σ, as a function of the thermodynamic forces, we have a paraboloid like that of Fig. 3.2, defined by

$$\sigma = A_{uu}X_u^2 + A_{ii}X_i^2 + (A_{ui} + A_{iu})X_uX_i \qquad (3.20)$$

According to Prigogine's theorem, if there are no restrictions on the forces, the system tends to the state of minimum entropy. This state, in the absence of restrictions, is the one that has zero entropy production.

In this case, the system tends toward the equilibrium state characterized by $X_u = 0$ and $X_i = 0$. However, if we externally hold one of the forces at a fixed value different from zero, the system will not achieve any equilibrium state but will tend toward the stationary state characterized by the minimum production of entropy compatible with the fixed condition (that is, the system will move along the parabola of Fig. 3.2 until it stops at point S, which is its minimum). The minimum condition in this case is that the derivative of σ with respect to X_i (we assume that X_u is held fixed) be zero. That is,

$$d\sigma/dX_i = 2A_{ii}X_i + (A_{ui} + A_{iu})X_u = 0 \tag{3.21}$$

Now then, if the Onsager relations (3.15) are satisfied, $A_{ui} = A_{iu}$ and the last expression can be written as

$$2(A_{ii}X_i + A_{iu}X_u) = 2J_i = 0$$

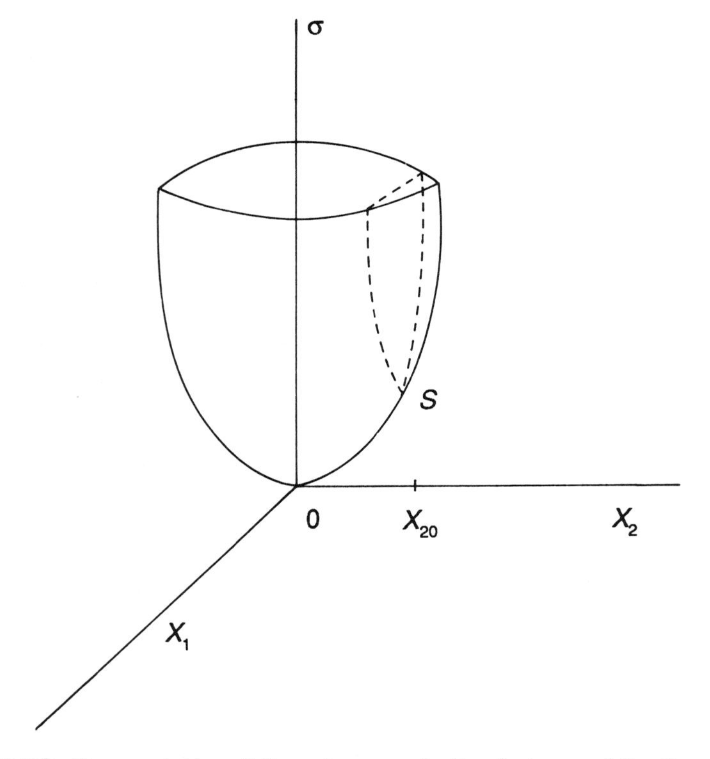

Figure 3.2 Representation of the entropy production in terms of the thermodynamic forces. If the system is allowed to evolve freely, it evolves toward the equilibrium state (the origin). If a fixed value X_{20} is maintained for the force X_2, the system evolves toward the stationary state characterized by minimum entropy production, in this case corresponding to point S.

since J_i, the material flux, according to (3.10) is

$$J_i = A_{ii}X_i + A_{iu}X_u$$

Therefore, the state of minimum entropy production is characterized by a zero value of the material flux; in other words, it is a stationary state with respect to material transport.

The theorem of minimum entropy production and some of its applications have generated extensive interest among biologists in the thermodynamics of irreversible processes. Various authors, including J. Wagensberg and collaborators at the University of Barcelona, have studied the growth of bacterial colonies calorimetrically. If we assume in principle, as a simplified hypothesis, that the heat dissipated by the system (measured by high-precision calorimeters) is directly related to the entropy produced by the same system due to its growth or differentiation, the calorimetric results indicate a stage of rapid increase in the entropy production, which goes to a maximum and then decreases to a minimum value (of course, positive), while the system remains in a stationary state. In this last state, the colony no longer grows; it has achieved its maximum capacity and is limited to maintaining itself. One interpretation of this phenomenon is to assume that on the global scale and in a stage in which large changes and growth no longer occur, the system can be described by means of linear laws and tends toward the minimum production of entropy. In fact, the resting stages of cells can often be interpreted as stationary states described by linear laws. Now then, the active phases of excitation or multiplication fall outside the range of purely linear behavior. We will study some examples of nonlinear phenomena in the last part of the book.

For greater clarity, let us present a concrete example of application of minimum entropy production in the calculation of certain geometric relations in the bronchial tree. The problem can be summarized in the following terms. The trachea is subdivided into two bronchi. Let us assume that each bronchus is subdivided in turn into two more, and so on. Let us call the different subdivisions "generations" (z). Thus the trachea is the zeroth generation, the bronchi are the first generation, their first subdivisions are the second generation, and so on.

In the first ten or twelve generations, we observe that the value of the diameter $d(z)$ of the bronchial tubes varies as

$$d(z) = d(0) \cdot 2^{-z/3}$$

in a way so that bronchioles of successive generations become narrower each time, according to a definite quantitative relation. The exponent $-z/3$ has various interpretations. One such interpretation is that it is the exponent which corresponds to minimum entropy production. Let us examine this interpretation.

The heat dissipated per unit time due to viscosity in each one of the bronchioles of the z generation can be written as

$$\text{heat } (z) = \text{hydrodynamic resistance } (z) \times [\text{volume flow } (z)]^2$$

where we indicate that heat is a function of each generation. This expression is analogous to Joule's law for electricity, with volume flow instead of electric current and hydrodynamic resistance instead of electrical resistance. The hydrodynamic resistance can be given by Poiseuille's law as

$$\text{hydrodynamic resistance } (z) = 8\eta l(z)/\pi r^4(z)$$

where $l(z)$ and $r(z)$ are the length and radius of the bronchioles, respectively, and η is the viscosity of air. We assume that the flow is laminar.

The entropy produced in each bronchiole per unit time and volume is

$$\sigma = \frac{1}{T}\frac{1}{\pi r^2(z)l(z)}\frac{8\eta l(z)}{\pi r^4}Q^2(z)$$

where $Q(z)$ is the volume flow for the considered bronchiole, T is the absolute temperature, and $\pi r^2 l$ is the volume of the bronchiole.

Now let us assume that

$$r(z) = r(0)2^{az}$$

The problem consists of determining the value of a. We can now write σ as

$$\sigma = \frac{8\eta}{\pi^2 r^6(z)T}\frac{Q^2(0)}{n^2(z)}$$

where we bear in mind that $Q(z) = Q(0)/n(z)$; that is, the volume flow for a bronchiole of the z generation is the volume flow in the trachea $Q(0)$ divided by the number of bronchioles of generation z, $n(z)$, which is equal to

$$n(z) = 2^z$$

Then we have

$$\sigma = 8\eta Q^2(0) \cdot 2^{-2z(1+3a)}/\pi^2 T$$

The condition of minimum entropy production is written as

$$\partial\sigma(z)/\partial Q(z) = 0$$

We introduce a symbol x defined as $x \equiv 2^{-2z}$. The derivative of σ can be rewritten as

$$\frac{\partial \sigma(z)}{\partial Q(z)} = \left|\frac{\partial \sigma}{\partial x}\right| \left|\frac{\partial x}{\partial Q}\right| = -\frac{8\eta}{\pi^2 T} Q^2(0)(1 + 3a)x^{3a} \frac{2x^{3/2}}{Q(0)}$$

This expression is set to zero when $Q(0) = 0$ (equilibrium case, no flow and no dissipation), or when the system is in a nonequilibrium stationary state with $Q(0) \neq 0$ and when $1 + 3a = 0$, that is, when $a = -1/3$. This is precisely the value of the exponent observed in practice.

We should warn that this formulation for the pulmonary system does not fit reality too well, since the flow is not strictly stationary but rather is pulsed. Therefore, the value of the flow must not be interpreted exactly like instantaneous flow, but rather like the mean square value of a pulsed flow. However, this situation is interesting because it shows that the theorem of minimum entropy production is not just a formal result, but on the contrary has quantitative predictive power.

3.8 CONCLUSIONS

The central quantity of this chapter is entropy production. In the first place, the second law, which states that the entropy production is positive, imposes restrictions on the phenomenological equations describing the dynamics of the system. Second, entropy production consistently defines the combination of thermodynamic forces conjugate to each combination of dissipative fluxes, an essential condition for using the Onsager relations between the phenomenological coefficients of the linear equations. Third, and within the framework of the linear theory, the fact that entropy production is minimal in stationary states allows us to sometimes arrive at quantitative predictions about some parameters of the system. These results have been explained in this chapter.

In addition, we mention that the production of entropy has other interesting aspects. On the one hand, its minimal character in stationary states makes it very useful for studying the stability of states. Second, entropy production is related to the decrease in efficiency of real processes in comparison with reversible processes, so that often minimizing entropy production helps us maximize the efficiency of processes. Finally, entropy production has been used by some authors to establish a link between physical time and biological time. In other words, entropy production might represent a certain measure of biological age. This last question, despite its interest, is still in a very speculative phase. Therefore, it would be premature to include it in an introductory book.

BIBLIOGRAPHY

de Groot, S. R., and P. Mazur, *Non-Equilibrium Thermodynamics*. Amsterdam: North Holland, 1962.

Glansdorff, P., and I. Prigogine, *Thermodynamics, Structure, Stability and Fluctuations*. New York: Wiley, 1971.

Kane, J. W., and M. M. Sternheim, *Physics*, 2nd ed. New York: Wiley, 1983.

APPLICATIONS OF THE LINEAR THEORY

4

Diffusion and Sedimentation

4.1 INTRODUCTION

One of the most important topics in biophysics is material transport. Cells import substances (nutrients, for example) which the metabolism reduces either to waste products or to substances that store energy. The waste products must be expelled and the substances that store energy must be distributed to the points where this energy is needed. The cell is found in a dynamic situation: it needs to feed regularly, even though it is at rest. Material transport is important not only at the cellular level, but also both at the tissue level and in transport of organisms in fluid media. Thus the study of material transport is important in physiology (cellular, animal, plant), in ecology, and even in molecular biology. In this chapter, we will study the phenomenon of isothermal diffusion in the absence of membranes, that is, for systems immersed in a fluid at constant temperature. In the following chapter, we will study material transport across passive membranes.

4.2 ENTROPY PRODUCTION. FICK'S LAW

In the last chapter, we obtained the entropy production in a continuous system. If the temperature and pressure are constant and only the concentration of the substances varies with position (x), then the entropy production (3.17) is reduced to

$$\sigma = -J_i(d\mu_i/dx)T \geq 0 \tag{4.1}$$

As we did in the last chapter for the case of thermal energy, let us assume that the flux J_i can be given by

$$J_i = -A_{ii}T^{-1}(d\mu_i/dx) \tag{4.2}$$

where, according to the second law, the coefficient A_{ii} is positive. In an ideal system, the chemical potential is related to the concentration by (2.20), and therefore for constant temperature and pressure we have

$$d\mu_i/dx = (RT/c_i)(dc_i/dx) \tag{4.3}$$

where c_i is the concentration (in moles per unit volume) of the substance i. By means of this expresion, (4.2) becomes

$$J_i = -(A_{ii}R/c_i)(dc_i/dx) = -D(dc_i/dx) \tag{4.4}$$

This relation defines D, which is called the *diffusion coefficient* and is of great importance in many problems in biophysics. This law

$$J_i = -D(dc_i/dx) \tag{4.5}$$

is called *Fick's law*, in honor of the German physiologist who discovered it in 1855. The time variation of the concentration can be given by

$$\partial c_i/\partial t = -dJ_i/dx = D\partial^2 c_i/\partial x^2 \tag{4.6}$$

This last equation is the famous diffusion equation, in which $c_i(t, x)$ is a function of two variables: time and position. This is another difference between continuous and discrete systems, where the flux J_i is related directly to the variation in c_i with respect to time, with no need for any spatial derivative. The time derivative of c_i is given as the difference between the flux which leaves at $x + dx$, $J_i(x + dx)$, from the flux which enters at x, $J_i(x)$, in order to thus obtain the variation in the number of molecules within the volume dV, that is,

$$\begin{aligned}(dc_i/dt)dV &= AJ_i(x) - AJ_i(x + dx) \\ &= AJ_i(x) - A[J_i(x) + (dJ_i/dx)dx] = -(dJ_i/dx)Adx\end{aligned}$$

which leads to (4.6), since $dV = Adx$ is the differential of the volume.

In our analysis, let us assume that the solvent is at rest and that there is only one solute. The diffusion flux will be its relative movement with respect to the solvent.

4.3 DIFFUSION IN A GRAVITATIONAL FIELD. EINSTEIN RELATION

We have seen in Eq. (2.22) that, in a gravitational field, the expression for the chemical potential is modified by inclusion of the gravitational potential energy per mole. The equilibrium condition for there to be no diffusion flow is

$$d\mu_i/dx = 0$$

where now we assume that x is the vertical axis. Keeping this condition and (2.22) in mind, we obtain

$$\frac{d\mu_i}{dx} = \frac{RT}{c_i}\frac{dc_i}{dx} + M_i'g = 0$$

where M'_i is the molar apparent mass of the substance i [see the definition of molar mass in Eq. (4.9)], g is the acceleration of gravity, R is the universal gas constant, and T is the absolute temperature. From the last expression we get

$$\frac{dc_i}{c_i} = -\frac{M'_i g}{RT}\,dx$$

Integrating this equation, we obtain

$$c_i(x) = c_i(0)\exp[-M'_i gx/RT] \tag{4.7}$$

where $c_i(0)$ is the concentration of i at the level which we take as the reference level, and $c_i(x)$ is the concentration at a height x above the reference level. Relation (4.7) gives the distribution of the concentrations over the height in an equilibrium situation.

Next, we look for this equilibrium distribution in a way that is less direct but which allows us to obtain a series of very useful results, since they relate the problem of diffusion to the problem of sedimentation and allow us to understand how the compromise between these two phenomena leads to the equilibrium condition (4.7).

First of all, let us recall that a small particle moving slowly in a fluid experiences a force of resistance to its movement (or drag force), proportional to its velocity,

$$\vec{F}_r = -\alpha\vec{v} \tag{4.8}$$

If the particle is a sphere of radius r and the fluid has viscosity η, the coefficient α, according to Stokes's law, is $\alpha = 6\pi\eta r$.

Second, let us recall that, due to Archimedes's principle, the apparent mass of a particle of true mass m and density ρ_0 in a fluid of density ρ is

$$m' = m[1 - (\rho/\rho_0)] \tag{4.9}$$

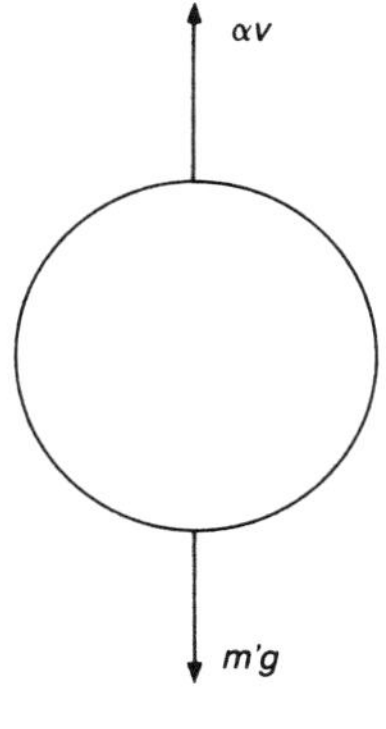

Figure 4.1 Sphere in a fluid subject to the gravitational force (apparent weight = true weight minus bouyant force) and to the hydrodynamic resistance force (drag force) proportional to the velocity.

where we bear in mind the buoyant force to which the particle is subjected (opposing gravity), equal to the weight of the displaced fluid.

Then we will be in a position to study the fall of a particle immersed in a fluid. Newton's second law is written as

$$m(dv/dt) = m'g - \alpha v \tag{4.10}$$

When the velocity is low, the resistance force is negligible and the particle accelerates. As the velocity increases, the resistance force increases until, for a certain velocity (the sedimentation velocity), the resistance force is equal to the apparent weight. Since now the sum of the forces is zero, the particle does not accelerate further and continues sedimentation at a uniform velocity given by the condition $dv/dt = 0$ in (4.10); that is,

$$v_{\text{sed}} = (m'/\alpha)g \tag{4.11}$$

In the particular case of spherical particles, we have

$$v_{\text{sed}} = \frac{(4/3)\pi r^3(\rho_0 - \rho)g}{6\pi\eta r} = (2/9)(r^2/\eta)(\rho_0 - \rho)g \tag{4.12}$$

This formula allows us to confirm that larger particles fall more rapidly, and that particles settle more slowly in more viscous fluids. Both observations are intuitive and are in accord with experience. In many cases, in ecology problems (sedimentation of plankton) it is necessary to use the turbulent viscosity instead of the molecular viscosity; the former is several thousand times greater than the latter.

Due to the sedimentation process, the particles accumulate at the bottom of the container. Then a concentration difference is created; this generates an upward diffusion flow which becomes more intense as the number of particles accumulating on the bottom increases, until the sedimentation flow is exactly counteracted. In this way we arrive at a stationary situation in which the descending sedimentation flow, given by

$$J_{\text{sed}} = cv_{\text{sed}}$$

is equal to the ascending diffusion flow, given by

$$J_{\text{dif}} = -D(dc/dx)$$

Equating both fluxes, we obtain an equation for the steady-state distribution of concentrations in equilibrium

$$cv_{\text{sed}} = -D(dc/dx) \tag{4.13}$$

This equation can be immediately integrated and leads to

$$c(x) = c(0)\exp[-v_{\text{sed}}x/D] \tag{4.14}$$

where, as in (4.7), $c(0)$ is the concentration at the reference height. Since the distributions (4.7) and (4.14) describe exactly the same physical situation, they must be identical. Comparing them and recalling (4.11), we make the identification

$$D = kT/\alpha \tag{4.15}$$

where k is the Boltzmann constant ($k = R/N_A$), the ideal gas constant divided by Avogadro's number. This relation was obtained by Einstein in 1905 (the same year in which he formulated the special theory of relativity). In order to clearly understand its importance, let us write it for the case of spherical particles. According to Stokes's law, we have

$$D = kT/6\pi\eta r \tag{4.16}$$

This equation shows how the diffusion coefficient D depends on the temperature, the solvent (the viscosity), and the solute (the particle radius). The information contained in this equation is easy to grasp intuitively. The larger the particles, the more it takes to move them and the more slowly they diffuse. The same thing happens as the viscosity of the solvent becomes greater, which hinders movement of the particles. On the other hand, a higher absolute temperature (related to thermal agitation) favors rapid diffusion. In general, the particles of interest in biophysics are macromolecules or microorganisms for which we do not know the radius or the mass (or even the shape!). In the case when they are more or less spherical, the Einstein relation (4.16) allows us to calculate the radius starting from measurement of the diffusion coefficients, the viscosity, and the absolute temperature.

We observe that if the molecules are globular and of constant density, their mass will be proportional to the volume, that is, to r^3. In other words, if we invert the relation we will find that the radius r is proportional to the cube root of the mass M of the molecules, and therefore, according to (4.16),

$$D \sim 1/M^{1/3}$$

Therefore, heavier molecules have smaller diffusion constants, as we can see in the entries in Tables 4.1 and 4.2.

TABLE 4.1 Diffusion Coefficients of Some Molecules in Water (20°C)

Molecule	M (g/mole)	D (cm²/s)
Oxygen	32	$1.0 \cdot 10^{-5}$
Urea	60	$1.1 \cdot 10^{-5}$
Glucose	180	$6.7 \cdot 10^{-6}$
Ribonuclease	13,683	$1.2 \cdot 10^{-6}$
Hemoglobin	68,000	$6.9 \cdot 10^{-7}$
DNA	6,000,000	$1.3 \cdot 10^{-8}$

TABLE 4.2 Diffusion Coefficients of Some Molecules in Air (20°C, 1 atm)

Molecule	D (cm^2/s)
Water	0.239
Oxygen	0.178
Carbon dioxide	0.139

The Einstein formula is modified when the solution is concentrated, since in this case there are hydrodynamic interactions between the different molecules, which do not move within a completely quiescent liquid but rather are altered by the movement of other molecules. The Einstein relation then takes on a form like the following:

$$D = (kT/\alpha)(1 + B_1c + B_2c^2)(1 + E_1c)^{-1} \quad (4.17)$$

In general, D decreases when c (the concentration) increases; frequently this decrease is linear.

$$D = D_0(1 - kc)$$

with D_0 as the diffusion constant at very low concentration, given by (4.15). If in addition there are other kinds of interactions between the molecules (electric, magnetic, and so on), the modification of the Einstein relation is much more extensive.

4.4 DIFFUSION AND BROWNIAN MOVEMENT

A particle of pollen (or any very small particle), suspended in a fluid, undergoes incessant erratic motion. This observation, made by the botanist Brown, has been a paradox for physicists for years. Is this perpetual motion? It was Einstein who gave the definitive explanation. Pure and simple, this movement is a visible manifestation of the molecular agitation which (according to the energy equipartition theorem) the particle in suspension shares with the molecules of the fluid, being struck continually by them in all directions and being pushed, decelerated, deflected, and accelerated.

On the basis of his study, Einstein was able to prove that, on the average, the deviation Δx of each particle with respect to its initial position satisfies the relation

$$\overline{\Delta x^2} = 2Dt \quad (4.18)$$

where D is the diffusion coefficient and t is the time. The horizontal bar

Figure 4.2 Trajectory of a particle in Brownian motion.

indicates averaging over the different Brownian particles. In order to see how extraordinary this formula is, let us recall that if the movement were rectilinear and uniform then the deviation or distance traveled would be proportional to the time, and not to the square root of the time as in (4.18). This decrease in the distance to time ratio is logical, since the molecule advances in an oscillatory manner, forward and backward, instead of moving always forward.

The information contained in this formula can be used in some examples as interesting as calculation of the velocity of blood circulation in pulmonary alveoli. With what velocity must the blood circulate for optimal exchange of gases (mainly oxygen and carbon dioxide) with the air in the lungs? If the blood moves too rapidly, there is no time for exchange of the two gases. If it moves too slowly, the result is an unnecessary loss of time.

In order to determine how long it takes for oxygen to diffuse into the blood, we recall that the alveolar wall is very thin (about 0.2 μm) and that the radius of the alveoli is on the order of 5 μm. Taking the value of D in Table 4.1 and using (4.18), we find that the time it takes for oxygen to diffuse into the blood is

$$t \approx \overline{\Delta x^2}/2D \approx r^2/2D = (5 \cdot 10^{-4})^2/2 \cdot 10^{-5}\ \mathrm{cm^2/s}$$
$$= 12 \cdot 10^{-3}\ \mathrm{s} \approx 0.01\ \mathrm{s}$$

Therefore the transit time for blood through a capillary in contact with the alveolus must be on the order of a hundredth of a second.

In practice, the velocity of blood in the capillaries is about 0.1 cm/s and the typical length of the capillary in the alveolus is 100 μm = 0.01

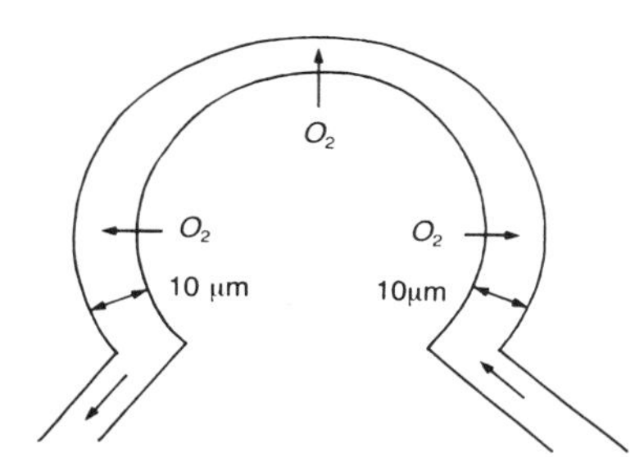

Figure 4.3 Capillary in the wall of a pulmonary alveolus.

cm, so the transit time is really 0.1 s. This is enough time for good exchange to occur, and is short enough to ensure good transport efficiency. We see then how, knowing the diffusion coefficient and a few elementary properties, we discover equilibria and unsuspected harmonies in the rhythm of physiological processes. It goes without saying that this type of calculation can be applied, for example, to the functioning of the kidneys and other organs.

4.5 DIFFUSION AND ULTRACENTRIFUGATION

The process of sedimentation in a gravitational field is extremely slow for very small particles. In order to accelerate this process and convert it into a technique of experimental interest, it is profitable to use centrifugal acceleration fields instead of gravity. Thus $\omega^2 R$ can replace the value of g (the acceleration of gravity), where R is the radius of gyration and ω is the angular velocity of rotation. A typical value of R is about 6 cm, which gives us a centrifugal acceleration

$$a_c = \omega^2 R = (0.24 s^2 f^2) g \tag{4.19}$$

where g is the acceleration of gravity and f is the number of revolutions per second.

When $f = 10$ rev/s, $\omega^2 R = 24g$, and if $f = 10^3$ rev/s, $\omega^2 R = 240{,}000g$. Thus, with machines that rotate sufficiently rapidly, we can achieve accelerations much greater than the acceleration of gravity; since the sedimentation velocity given by (4.12) is proportional to the acceleration, the sedimentation process occurs much more rapidly and more efficiently. Ultracentrifugation has been developed as an experimental technique since 1920, especially by the work of Svedberg (Nobel Prize in Chemistry) and his group. Today, it is used as a technique for analysis and separation of different substances, due to the fact that the sedimentation velocity depends [as we have seen in (4.12)] on the particle dimensions. Consequently, by acquiring different velocities, the particles will be separated and stratified during the process.

The sedimentation velocity is now, according to (4.9), (4.11), and (4.15),

$$v_{\text{sed}} = (D/kT)m[1 - (\rho/\rho_0)]\omega^2 R \tag{4.20}$$

where we recall that, according to the Einstein relation (4.16), $D/kT = 1/\alpha = 1/6\pi\eta r$, ρ is the density of the liquid, and ρ_0 is the density of the particle, and the centrifugal acceleration $a_c = \omega^2 R$ is used instead of g in (4.11).

We generally define the *sedimentation coefficient* as

$$S \equiv v_{\text{sed}}/\omega^2 R = (D/kT)[1 - \rho/\rho_0)] \tag{4.21}$$

As we can see, S depends only on the settling molecule, the temperature, and the density of the solution. If we multiply and divide by Avogadro's number, we obtain

$$S = (D/RT)[1 - (\rho/\rho_0)]M \tag{4.22}$$

where $M = m\mathcal{N}_A$ is the molar mass (m is the molecular mass) and $R = k\mathcal{N}_A$ is the ideal gas constant. We also see that S has the dimensions of time, with typical values on the order of 10^{-13} s, or $M \approx 10^5$ g/mole, $D \approx 10^{-7}$ cm^2/s, $T \approx 300$ K, $R = 8.31 \cdot 10^7$ ergs/mole $\cdot$ K, and $1 - (\rho/\rho_0) \approx 0.25$.

The value of $S = 10^{-13}$ s is called the svedberg and is taken as the unit of sedimentation coefficients. A sedimentation coefficient of 5.3 svedberg then means $S = 5.3 \cdot 10^{-13}$ s. If $S = 3.6$ svedberg and if $\omega^2 R = 10^5$ g, then the sedimentation velocity is about $3.6 \cdot 10^{-5}$ cm/s and the centrifuge should rotate for an hour in order to produce a 10^{-1}-cm displacement of the molecules. This gives an idea of the extreme slowness of diffusion and sedimentation processes, even in centrifuges.

Often (4.22) is used to calculate the molar mass starting from the diffusion coefficient and the densities and the sedimentation coefficient, since

$$M = (RT/D)S\rho_0(\rho_0 - \rho)^{-1} \tag{4.23}$$

The density of the particles is determined approximately as the reciprocal of the specific volume of the solute molecules, which can be obtained by measuring the volume increment for a dilute solution when a gram of dry solute is added to it. This is generally the more difficult factor to determine, while the others (D, T, S, and ρ) are obtained somewhat more immediately. In Table 4.3, we have collected some data related to the problem of sedimentation in ultracentrifuges.

Another procedure for obtaining the molecular mass involves examining the density distribution as a function of the radius of gyration r_g in the centrifuge. Introducing this parameter into (4.13) and inte-

TABLE 4.3 Some Parameters Involved in the Study of Centrifugation

	$1/\rho_0$ (cm^3/g)	S (svedbergs)	D ($10^{-7}cm^2/s$)	M (g/mole)
Myoglobin	0.741	2.04	11.3	16,900
Lactoglobulin	0.751	3.12	7.3	41,500
Egg albumin	0.749	3.55	7.8	44,000
Hemoglobin	0.749	4.41	6.3	68,000

grating, we obtain the following for the equilibrium distribution:

$$c(r_{g1})/c(r_{g2}) = \exp[-M'\omega^2(r_{g1}^2 - r_{g2}^2)/2RT]$$

from which it is easy to obtain

$$M = \frac{2RT\rho_0}{\omega^2(r_{g2}^2 - r_{g1}^2)(\rho_0 - \rho)} \ln \left|\frac{c(r_{g1})}{c(r_{g2})}\right| \qquad (4.24)$$

This expression for M is more precise, since it does not involve the diffusion coefficient D or the shape of the molecules. On the other hand, it is inconvenient to have to wait a long time to ensure that equilibrium has been achieved. In the meantime, expression (4.23) is applicable immediately, without any need to have achieved equilibrium.

BIBLIOGRAPHY

BERG, H. C., *Random Walks in Biology.* Princeton, New Jersey: Princeton University Press, 1983.

KATCHALSKY, A., and P. F. CURRAN, *Nonequilibrium Thermodynamics in Biophysics.* Cambridge, Massachusetts: Harvard University Press, 1965.

OLANDER, D. R., "The gas centrifuge," *Scientific American* 239, No. 2 (1978), 37–43.

VILLARS, F. M. H., and G. B. BENEDEK, *Physics With Illustrative Examples From Medicine and Biology,* Vol. 2. Reading, Massachusetts: Addison-Wesley, 1974.

VOGEL, S., *Life in Moving Fluids.* Princeton, New Jersey: Princeton University Press, 1985.

5

Passive Transport Across Membranes

5.1 INTRODUCTION

Membranes play an important role in biology at both the cellular and the histological level, and there are monographs dedicated exclusively to this subject. In biological membranes there are various possible types of transport: purely passive transport, due to diffusion or a pressure gradient; facilitated transport, and active transport. The last two types are, of course, the most interesting from a biological point of view. We devote Chapter 7 to studying active transport, while in this chapter we study strictly passive transport. Our specific objective is to describe the permeable membrane by means of the minimum number of parameters that characterize it completely. We will restrict ourselves to a single solute and to a single solvent, and to nonelectrolytic solutes. Even so, the subject is far from uninteresting, since it establishes the fundamental concepts for study of the phenomenon of transport across membranes.

5.2 ENTROPY PRODUCTION

According to the treatment in Chapter 3, in an isothermal system in which the only irreversible process is material transport, the entropy production takes on the form

$$\sigma = -\Sigma_i J_i T^{-1} \Delta\mu_i \tag{5.1}$$

If we consider a single solute in a single solvent, the last expression is reduced to

$$-T\sigma = J_s \Delta\mu_s + J_w \Delta\mu_w \tag{5.2}$$

where J_s is the solute flux and J_w is the solvent flux. We observe that the equations are those for a discrete problem, since we consider that the membrane divides the system into two subsystems which, under good agitation conditions, can be considered as practically homogeneous.

In principle, we can work directly with these two fluxes. But much more often we work in terms of the total volume flux across the membrane, J_v, and the velocity of the solute relative to the solvent, J_d, defined respectively as

$$J_v = J_w\overline{V}_w + J_s\overline{V}_s \tag{5.3}$$

$$J_d = v_s - v_w \tag{5.4}$$

where $\overline{V}_s$ and $\overline{V}_w$ are the partial volumes of the solute and the solvent, and v_s and v_w are their respective velocities.

Since the chemical potential appears in Eq. (5.2), it is of interest to study it in detail in a more convenient form. Starting from the Gibbs equation (2.25)

$$dG = -SdT + Vdp + \mu_s dN_s + \mu_w dN_w$$

we have

$$\left.\frac{\partial G}{\partial N_w}\right|_{T,p,N_s} = \mu_w$$

$$\left.\frac{\partial G}{\partial p}\right|_{T,N_w,N_s} = V$$

The equality of the second derivatives leads to the relation

$$\left.\frac{\partial V}{\partial N_w}\right|_{T,p,N_s} = \left.\frac{\partial \mu_w}{\partial p}\right|_{T,N_s,N_w}$$

Given the relationship between the total volume V and the partial molar volume of the solute and the solvent,

$$V = N_w\overline{V}_w + N_s\overline{V}_s$$

where N_w and N_s are, respectively, the number of moles of solvent and solute, we obtain

$$V_w = \left.\frac{\partial \mu_w}{\partial p}\right|_{T,N_s,N_w}$$

Integrating this equation, we obtain the explicit dependence of the chemical potential on p and some undetermined function of temperature and concentration:

$$\mu_w = \overline{V}_w p + \text{function } (T, c)$$

which we finally write as

$$\mu_w = \mu_w^0(T) + \overline{V}_w p + \mu_w^c \tag{5.5}$$

where

$$\mu_w^c = RT \ln c_w$$

In order to go from one type of fluxes to others and in order to see the convenience of this transformation, let us assume a membrane which, as we see in Fig. 5.1, separates two regions with different concentrations.

Let us calculate $\Delta\mu_w$, where Δ indicates the difference in the values of the corresponding parameter (in this case, the chemical potential of the solvent) on both sides of the membrane.

$$\Delta\mu_w = (\mu_w)_0 - (\mu_w)_{\Delta x}$$

If we use Eq. (5.5), the last equation can be described as

$$\Delta\mu_w = \overline{V}_w \Delta p + (\Delta\mu_w^c)_0 - (\Delta\mu_w^c)_{\Delta x} \tag{5.6}$$

and also as

$$\Delta\mu_w = \overline{V}_w \Delta p - \overline{V}_w \Delta\Pi \tag{5.7}$$

where Π is the osmotic pressure.

Let us recall that the osmotic pressure is the difference in pressure that must exist between the two compartments separated by a semipermeable membrane in order to have no net flow of solvent, that is, in order for the chemical potential of the solvent to be the same in the two compartments. As we can see in (5.7), when the pressure difference is equal to the difference in the osmotic pressures, the chemical potential of the solvent is in fact the same on the right as on the left of the membrane, since then $\Delta\mu_w = 0$.

Analogously, for the solute we have

$$\Delta\mu_s = \overline{V}_s \Delta p + \Delta\mu_s^c \tag{5.8}$$

where $\Delta\mu_s^c$ is the part of the chemical potential that depends on the concentration. In this case, it is not possible to directly relate $\Delta\mu_s^c$ to the osmotic pressure, in which we refer only to the chemical potential of the solvent and not that of the solute. For more convenience in writing the expressions, let us define the average concentration of the solute $\overline{c}_s$ as

$$\Delta\mu_s^c \equiv (1/\overline{c}_s)\Delta\Pi \tag{5.9}$$

where Π is the osmotic pressure defined by relation (5.7). Then this defi-

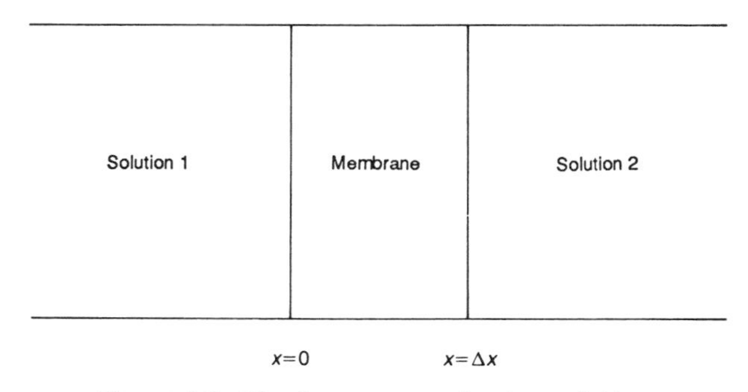

Figure 5.1 Membrane separating two solutions.

nition can be written

$$\Delta\mu_s = \overline{V}_s\Delta p + (1/\bar{c}_s)\Delta\Pi \quad (5.10)$$

The significance of relation (5.9) can be understood if we recall that, under ideal conditions, $\Delta\Pi = RT(c_{s1} - c_{s2})$ and that $\Delta\mu_s^c = RT(\ln c_{s1} - \ln c_{s2})$ and that therefore

$$\bar{c}_s = (c_{s1} - c_{s1})(\ln c_{s1} - \ln c_{s2})^{-1}$$

When $c_{s1}/c_{s2} = x \approx 1$, we can expand $\ln x = 2(x - 1)/(x + 1)$, with which the last expression is reduced to

$$\bar{c}_s = (1/2)(c_{s1} + c_{s2})$$

which is the average concentration or the mean of the concentrations to the right and to the left of the membrane. Under nonideal circumstances, $\bar{c}_s$ is not exactly the arithmetic mean of the concentrations.

Introducing the explicit expressions (5.7) and (5.10) into (5.2), we have

$$-T\sigma = (J_w\overline{V}_w + J_s\overline{V}_s)\Delta p + (J_s/\bar{c}_s - \overline{V}_wJ_w)\Delta\Pi \quad (5.11)$$

However, if $c_w\overline{V}_w \approx 1$, that is, if the volume fraction of water is practically unity (in other words, in the case of high degree of dilution), we have

$$J_s/\bar{c}_s - J_w\overline{V}_w \approx \bar{c}_sv_s/\bar{c}_s - \bar{c}_wv_w/\bar{c}_w = v_s - v_w \quad (5.12)$$

In this equality, we have considered that

$$J_w = \bar{c}_wv_w \quad \text{and} \quad J_s = \bar{c}_sv_s$$

that is, that in general terms, the flux is the product of the concentration times the velocity.

After all these transformations, we obtain the following expression for the entropy production in terms of J_v and J_d defined in (5.3) and (5.4)

$$-T\sigma = J_v\Delta p + J_d\Delta\Pi \quad (5.13)$$

Even though this section has been basically formal, we have felt compelled to write it in order to justify the treatment that appears in the rest of the chapter.

5.3 PHENOMENOLOGICAL EQUATIONS

The phenomenological equations that describe the response of the membrane, in accord with the concepts of linear thermodynamics which we have presented earlier, are

$$J_v = L_p\Delta p + L_{pd}\Delta\Pi \quad (5.14)$$

$$J_d = L_{dp}\Delta p + L_d\Delta\Pi \quad (5.15)$$

and the Onsager reciprocity relations establish that $L_{pd} = L_{dp}$. In this case, we see an example of the usefulness of these relations: the membrane, which in principle would be characterized by four parameters, needs only three parameters for its characterization. The significance of the phenomenological coefficients can be explored by examining various different physical situations. It is advisable to emphasize at this point the absolute necessity of writing the equations of the fluxes in terms of their respective conjugate forces in order to be able to use the Onsager reciprocity relations. For example, it would be incorrect to assume the symmetry of the coefficients if we express J_w and J_s in terms of Δp and $\Delta\Pi$, or J_v and J_d in terms of $\Delta\mu_s$ and $\Delta\mu_w$. This is why we have carried out the formal transformations of the last section in such detail, in order to be in a position to use the fluxes and forces that are as close as possible to the parameters of direct experimental interest.

Let us consider a situation in which the solute concentration is the same on both sides of the membrane and therefore $\Delta\Pi = 0$. If in spite of this there exists a hydrostatic pressure difference Δp between the two sides of the membrane, we will have a flux J_v that is a linear function of Δp. The proportionality constant relating both quantities is the *mechanical filtration coefficient* L_p, which represents the velocity of the fluid per unit pressure difference between the two sides of the membrane. A diffusion flux can also be produced by this pressure difference Δp, in such a way that the relation between J_d and Δp is adjusted by a coefficient L_{dp}, which is a measure of the ultrafiltration properties of the membrane. This coefficient, called the *ultrafiltration coefficient*, accounts for movement of the solute with respect to the solvent induced by a mechanical pressure. This is a well-known phenomenon in colloid chemistry. However, there are other possible physical situations where there are different solute concentrations ($\Delta\Pi \neq 0$) and in turn $\Delta p = 0$. Of course, the osmotic pressure difference produces a diffusion flux characterized by the coefficient L_d or the *permeability coefficient*, which indicates the movement of the solute with respect to the solvent due to the inequality of concentrations on both sides of the membrane. In addition, there is another phenomenon implied by Eq. (5.14) which involves a volume flux due to different osmotic pressures. This coefficient, L_{pd}, which relates the flux $(J_v)_{\Delta p=0}$ to $\Delta\Pi$, is called the *coefficient of osmotic flow*.

The introduction of cross coefficients is not arbitrary, but rather is imposed by the nature of the phenomena that occur in the membrane. For example, if we were to ignore L_{pd} in (5.14), we would lose the description of one of the most interesting functions of the membrane: selectivity. The importance of the cross coefficient L_{pd} can be revealed in the measurement of the osmotic pressure between two solutions separated by a membrane permeable to the solvent. The experiment is outlined in Fig. 5.3.

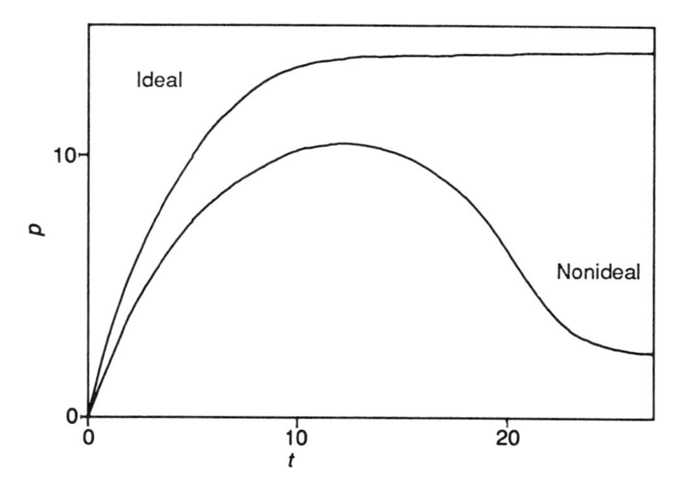

Figure 5.2 Hydrostatic pressure difference between two solutions separated by a semipermeable membrane. If the membrane is ideal (it does not allow the solute to pass), then the pressure difference increases until a certain fixed pressure is reached (the osmotic pressure). If the membrane is not ideal (that is, it allows passage of some of the solute), then the pressure difference increases until the osmotic pressure is reached; but then, as solute is lost from one solution to the other, the pressure difference slowly decreases until the pressure is the same in both solutions.

The height of the liquid in the capillary gives a measure of the change in volume and the pressure difference in "equilibrium." Generally we assume that the equilibrium state has been reached when $J_v = 0$. Then Δp measures the osmotic pressure difference between the two solutions.

As we can easily see starting from (5.14), this assumption is not always correct since, if $J_v = 0$, we have

$$(\Delta p)_{J_v=0} = -\frac{L_{pd}}{L_p}\Delta\Pi \qquad (5.16)$$

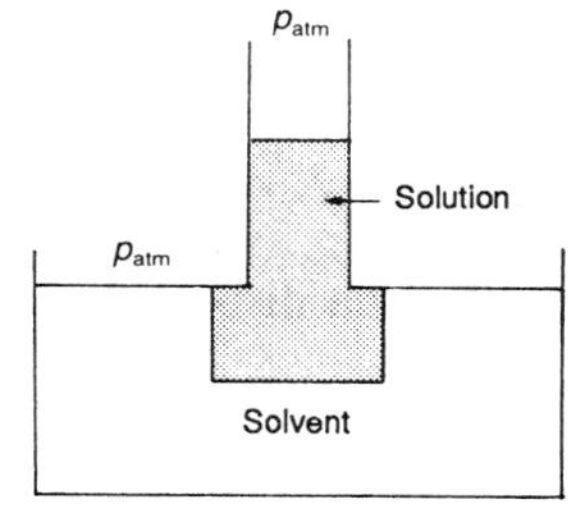

Figure 5.3 Measurement of the osmotic pressure between two solutions separated by a semipermeable membrane.

and only if $-L_{pd} = L_p$ do we have

$$(\Delta p)_{J_v=0} = \Delta\Pi$$

This condition will be satisfied only for ideal semipermeable membranes, which block the transport of solute no matter what the value of Δp and $\Delta\Pi$. In fact, if we evaluate J_s by summing J_v and J_d according to the arguments for (5.12), that is,

$$J_d + J_v = (J_s/\bar{c}_s) - J_w\overline{V}_w + J_w\overline{V}_w + J_s\overline{V}_s = (1 + \bar{c}_s\overline{V}_s)J_s/\bar{c}_s$$

and if we assume, as in (5.12), a very dilute solution,

$$\bar{c}_s\overline{V}_s \ll 1$$

we have

$$J_s/\bar{c}_s = (L_p + L_{pd})\Delta p + (L_{dp} + L_d)\Delta\Pi \qquad (5.17)$$

In order to obtain $J_s = 0$ for any value of Δp and $\Delta\Pi$, the following relations must be satisfied:

$$L_p = -L_{pd}, \qquad L_{dp} = -L_d, \qquad L_p = L_d = -L_{pd} \qquad (5.18)$$

and therefore

$$(\Delta p)_{J_v=0} = \Delta\Pi$$

If this is not the case, and the membrane allows solute to pass,

$$-L_{pd}/L_p < 1 \qquad (5.19)$$

For a selective membrane, like a porous glass filter, L_{pd} tends toward zero and therefore $-L_{pd}/L_p$ also tends toward zero. Staverman has called this quotient the *reflection coefficient* r, indicating, when $r = 1$, that all the solute is reflected at the membrane (ideal membrane) so that it cannot cross the membrane. When $r < 1$, on the other hand, some of the solute is reflected and the rest crosses the membrane.

When $r = 0$, the membrane is completely permeable. This is proved by considering once more the situation $\Delta\Pi = 0$ and calculating

$$\left[\frac{J_d}{J_v}\right]_{\Delta\Pi=0} = \frac{L_{pd}}{L_p} = -r \qquad (5.20)$$

Introducing the values of J_d and J_v in terms of the velocities, we obtain

$$\frac{v_s - v_w}{v_w} = -r, \quad \text{which implies} \quad \left[\frac{v_s}{v_w}\right]_{\Delta\Pi=0} = 1 - r$$

Thus the last equation shows that for an ideal semipermeable membrane, $r = 1$ and $v_s = 0$, so that none of the solute molecules can cross the membrane. When r decreases, v_s increases; and in the limit when $r = 0$, $v_s = v_w$; that is, the membrane is not selective and both the solute and the solvent move with the same velocity. It may also happen that r is negative, and in this case $1 - r > 1$. This means that $v_s > v_w$, a case known as negative anomalous osmosis. The transport of electrolytes across charged membranes, where we can also observe this phenomenon, is also of special interest.

The introduction of r in Eq. (5.12) entails

$$J_v = L_p(\Delta p - r\Delta\Pi) \tag{5.21}$$

On the other hand, starting from (5.17) and (5.21), the equation for the solute flux can be expressed as

$$J_s = \bar{c}_s(1 - r)J_v + \omega\Delta\Pi \tag{5.22}$$

where ω is the *solute permeability coefficient* and is given by the expression

$$\omega = (\bar{c}_s/L_p)(L_pL_d - L_{pd}^2)$$

For ideal membranes, $r = 1$ and $\omega = 0$. For nonselective membranes,

$$L_{pd} = 0 \quad \text{and} \quad \omega = \bar{c}_s L_d$$

This is a characteristic parameter both in synthetic membranes and in natural membranes. The thermodynamics of irreversible processes allows us to understand why three coefficients are sufficient to describe the transport of substances across a membrane and which relations exist among these coefficients. In Table 5.1, we give some parameters for two characteristic biological membranes.

TABLE 5.1 Phenomenological Coefficients of Different Membranes for Some Solutes in Water

Membrane	Solute	ω (10^{-15} moles/dynes·s)	r	L_p (10^{-11} cm^3/dynes·s)
Nitella translucens	Methanol	11	0.5	1.1
	Urea	0.008	1	—
Human red blood cell	Methanol	122	—	—
	Urea	17	0.62	0.92

TABLE 5.2 Reflection Coefficients for Some Solutes and Membranes

Solute	Nitella flexilis	Nitella translucens	Valonia utricalaris
Urea	0.91	1	0.76
Sucrose	0.97	—	1
Isopropanol	0.35	0.27	—
Ethanol	0.34	0.29	—
Methanol	0.31	0.25	—

5.4 MICROSCOPIC INTERPRETATION OF THE PHENOMENOLOGICAL COEFFICIENTS

In order to better understand the significance of these coefficients, let us imagine an extremely simplified model of the membrane, considering it as permeated by parallel cylindrical pores of radius a. Let us assume that there are N_p pores per unit surface area of the membrane. *Poiseuille's law*, well-known in elementary physics, establishes that the volumetric flow rate or volume flow (the volume which flows per unit time) of a liquid of viscosity η in a cylinder of radius a and length l, between the ends of which a pressure difference Δp is maintained, is equal to

$$Q = (\pi a^4/8\eta l)\Delta p \tag{5.23}$$

Thus the flux across the membrane per unit area and time and under the action of a pressure difference Δp is

$$J_v = N_p(\pi a^4/8\eta l)\Delta p \tag{5.24}$$

where we assume that the length of the tube coincides with the thickness l of the membrane. Comparing with (5.21), we find that the filtration coefficient can be expressed as

$$L_p = N_p(\pi a^4/8\eta l) \tag{5.25}$$

Let us now assume that there is no pressure difference between the two sides of the membrane, but there is a concentration difference and the solute can diffuse through the solvent and through the pores. If the solution is ideal, according to van't Hoff's law,

$$\Delta\Pi = RT\Delta c_s$$

and then we have

$$J_s = \omega\Delta\Pi = RT\omega\Delta c_s \tag{5.26}$$

According to Fick's law of diffusion, the diffusion flux in each pore of transverse area πa^2 and length l (the thickness of the membrane) is

$$J_s \text{ (1 pore)} = D(\pi a^2/l)\Delta c_s$$

where D is the diffusion coefficient. Identifying (5.26) with the total flux per unit area obtained according to Fick's law,

$$J_s = N_p D(\pi a^2/l)\Delta c_s \tag{5.27}$$

we obtain the following expression for the coefficient ω

$$\omega = N_p D \pi a^2/RTl \tag{5.28}$$

If some fraction r' of the solute is reflected, in general we obtain

$$\omega = (N_p D/RT)(\pi a^2/l)(1 - r') \tag{5.29}$$

Finally, various expressions have been proposed for the reflection coefficient r' from a hydrodynamic point of view. The simplest of these is

$$1 - r' = [1 - (r_s/a)]^2 \tag{5.30}$$

where r_s is the radius of the solute molecules and a is the radius of the pores. If $r_s > a$, we assume that $r' = 1$, because then, since the solute molecules do not fit into the pores, all the solute is reflected. This equation represents the idea that the effective area of the pores is $\pi(a - r_s)^2$, and therefore

$$\frac{\text{effective area}}{\text{true area}} = \frac{(a - r_s)^2}{a^2} = [1 - (r_s/a)]^2$$

In reality, the expression is somewhat more complicated when we consider a series of factors of a hydrodynamic nature. Table 5.3 includes some values of r' and ωRT as a function of the molecular radius, where we see how the reflection coefficient decreases as the molecular radius increases.

TABLE 5.3 Coefficients r' and ωRT for an Artificial Membrane with $a = 66 \cdot 10^{-10}$ m

Solute	Molecular radius ($\cdot 10^{-10}$ m)	r'	ωRT ($\cdot 10^{-2}$ m/s)
Urea	2.04	0.86	7.09
Glucose	4.44	0.82	3.42
Dextrin	8.98	0.55	1.17

5.5 APPLICATION: CONCENTRATION EQUALIZATION TIME

As a very simple application of the permeability coefficient, let us calculate the concentration equalization time between two zones separated by a membrane. This situation is of special interest in natural and artificial kidneys, in lungs, in cellular transport, and in many other common situations in biological problems.

Let us assume that we have two containers of volume V_1 and V_2 separated by a permeable membrane. The solute concentration on each side of the container is initially c_1 and c_2 with $c_1 > c_2$. Let us ask how much time it takes to equalize the concentrations in the two containers.

The decrease and increase in the number of moles of solute in containers 1 and 2 are given as a function of the area A of the membrane and the solute flux J_s from 1 to 2 by

$$dN_i/dt = -J_sA, \qquad dN_2/dt = J_sA \tag{5.31}$$

But $N_i = V_ic_i$ and $J_s = \omega RT(c_1 - c_2)$. Then

$$V_1(dc_1/dt) = -A\omega RT(c_1 - c_2)$$

$$V_2(dc_2/dt) = A\omega RT(c_1 - c_2)$$

Subtracting one expression from the other, we obtain

$$d(c_1 - c_2)/dt = -A\omega RT(V_1^{-1} + V_2^{-1})(c_1 - c_2) \tag{5.32}$$

The solution to this equation for $\Delta c = c_1 - c_2$ is

$$\Delta c(t) = \Delta c(0)\exp(-t/t_0) \tag{5.33}$$

where t_0 is defined, using (5.32), as

$$t_0 = (A\omega RT)^{-1}(V_1V_2)(V_1 + V_2)^{-1} \tag{5.34}$$

If we assume that $V_1 \gg V_2$ (the outer volume is much greater than the inner volume), we have a simplified expression for t_0:

$$t_0 \approx V_2\,(A\omega RT)^{-1} \tag{5.35}$$

Strictly speaking, equalization of concentrations is a process which, in the limit, is completed in an infinite time. In practice, however, after some time has passed (such as $t = 4t_0$), the concentration difference between the two containers will be virtually imperceptible and therefore the time t_0 can be considered as the characteristic time of the process of concentration equalization.

The usefulness of these calculations is clear, for example, in the design of artificial kidneys and hearts. We are able to know how long dialysis should continue for a patient based on the characteristics of the exchange membrane of the kidney. If we consider that the volume of interstitial fluid of an individual is about 40 liters, the area of the membrane of the kidney is about 2 m^2, and its permeability is $\omega RT = 5.5 \cdot 10^{-4}$ cm/s (for urea), then we obtain

$$t_0 = 3.5 \cdot 10^3 \text{ s} \approx 1 \text{ hr}$$

For uric acid, t_0 is twice this value, since the permeability is approximately half. At the end of four hours of operation, the patient would have eliminated 98 percent of the urea and 87 percent of the uric acid. The practical value of these results is obvious enough that we do not need to comment further.

BIBLIOGRAPHY

KATCHALSKY, A., and P. F. CURRAN, *Nonequilibrium Thermodynamics in Biophysics*. Cambridge, Massachusetts: Harvard University Press, 1965.

VILLARS, F. M. H., and G. B. BENEDEK, *Physics With Illustrative Examples from Medicine and Biology*, Vol. 2. Reading, Massachusetts: Addison-Wesley, 1974.

6

Coupled Chemical Reactions. Stoichiometry and Efficiency

6.1 INTRODUCTION

In addition to transport, the most frequently encountered processes in biological systems are those related to the chemical reactions of metabolism. The aspects of equilibria of chemical reactions (the equilibrium constant) which are the subject of elementary chemistry have been reviewed in Chapter 2. Their kinetic aspects can be studied from the standpoint of chemical kinetics, a well developed discipline. What can the thermodynamics of irreversible processes say that is new or unpublished? The most interesting and most typical aspects, as we will see, are the requirements regarding the apparent stoichiometry of two partially coupled reactions, and the study of the efficiency of such reactions as limited by the constraints of the second law of thermodynamics. These points of view concerning chemical reactions can be handled only within nonequilibrium thermodynamics.

6.2 EXTENT OF REACTION. AFFINITY

In Chapter 2, we saw that the Gibbs equation [Eq. (2.14)] for a multicomponent system takes on the form

$$dS = T^{-1}dU + pT^{-1}dV - \Sigma_i(\mu_i/T)dN_i \tag{6.1}$$

where N_i is the number of moles of species i and μ_i is its chemical potential, respectively. In studying chemical equilibrium in Section 2.6, we maintained that in each chemical reaction the variations in N_i are not independent, but rather are interrelated by the stoichiometry of the reaction. In the case of different reactions (which we distinguish by the subscript j), each one will be described by a chemical equation of the type

$$\Sigma_i \nu_{ij} x_i = 0 \tag{6.2}$$

Here ν_{ij} indicates the stoichiometric coefficient of the ith species (represented by x_i) in the jth reaction. We recall that $\nu_{ij} > 0$ for the products and $\nu_{ij} < 0$ for the reactants. Thus, as in (2.32), for each reaction we can define the extent of reaction $d\xi_j$ as

$$\frac{dN_{ij}}{\nu_{ij}} = d\xi_j \tag{6.3}$$

where dN_{ij} is the variation in N_i in the jth reaction.

In terms of the extent of reaction, we can describe the Gibbs equation (6.1) in the form

$$dS = T^{-1}dU + pT^{-1}dV - \Sigma_{ij}(\mu_i/T)\nu_{ij}d\xi_j \tag{6.4}$$

in which we have recalled that $dN_i = \Sigma_j dN_{ij}$, that is, that the variation in the total number of moles N_i is the sum of the variations in each and every reaction of the system.

It is also appropriate to define the *affinity* of each jth reaction as

$$A_j = -\Sigma_i \nu_{ij}\mu_i \tag{6.5}$$

The affinity depends on the concentrations, since the chemical potentials depend on the latter. For this reason, often when we speak of changing the affinities we refer to changes in the concentrations of the chemical species of the system. In terms of the affinities, (6.4) takes on the form

$$dS = T^{-1}dU + pT^{-1}dV - T^{-1}\Sigma_j A_j d\xi_j \tag{6.6}$$

In an isolated system in which U and V do not vary, $dU = dV = 0$, the entropy production can be given simply by

$$dS/dt = T^{-1}\Sigma_j A_j(d\xi_j/dt) \tag{6.7}$$

Let us consider the reaction rates (described by $J_i = d\xi_j/dt$) as the thermodynamic fluxes, and the respective affinities as the thermodynamic forces.

Thus the second law of thermodynamics will be expressed as

$$T\sigma = \Sigma_j A_j J_j \geq 0 \tag{6.8}$$

We observe that if there is only one reaction,

$$T\sigma = AJ \geq 0$$

that is, if $A > 0$, J must be positive; and if $A < 0$, we have $J < 0$. In order to have equilibrium, that is, in order for the reaction not to evolve in one direction or the other, we must have $A = 0$. This can be seen more clearly if we suppose, as in the general scheme of Chapter 3, a linear constitutive law

$$J = \alpha A \tag{6.9}$$

where α is a coefficient that can depend on temperature but not on the concentrations.

Thus, in (6.9) we see that the equilibrium condition (the situation in which the reaction does not evolve in any direction and therefore, with zero rate, $J = 0$) is that $A = 0$. If $A > 0$, then $J > 0$: the reactants decrease and the products increase—in other words, the reactants have an affinity for reacting. The equilibrium condition $A = 0$, according to the definition (6.5) for A, leads us to

$$A = -\Sigma\mu_i\nu_i = 0 \tag{6.10}$$

which is precisely the equilibrium condition (2.34) derived in Chapter 2. Then we recover the results of equilibrium thermodynamics, but with the advantage of having additional information about the rate at which equilibrium is achieved.

It is interesting to compare the constitutive law (6.9) with the results of the law of mass action of chemical kinetics. According to this law, the rate of a reaction is proportional to the product of the concentrations of the reactants (in the forward direction) and the products (in the backward direction). Thus, from left to right, the velocity w is

$$w = K\Pi_i c_i^{\nu_i}$$

and from right to left,

$$w' = K'\Pi_i c_i'^{\nu_i'}$$

where c refers to the reactants and c' refers to the products. The net reaction rate will then be

$$J = w - w' = w(1 - w'/w) = w[1 - \Pi_i c_i^{\nu_i}/K_c]$$

where K_c is the ratio of the two rate constants ($K_c = K/K'$), and can be calculated starting from the equilibrium concentrations, when $J = 0$. For a gas, we have

$$J = w[1 - \exp(-A/RT)] \tag{6.11}$$

The relation between reaction rate and affinity is not linear. It will be linear when A is much less than RT, since then we can expand the exponential in a series and we can neglect the terms higher than the first order, that is,

$$J = (w/RT)A + \text{terms of order } (A/RT)^2 \approx wA/RT \tag{6.12}$$

Comparing this equation with (6.9), we can identify the coefficient α with w/RT, a familiar parameter in chemical kinetics. In addition, we see that in general, the linear terms are not sufficient to account for various processes. Even so, many situations of biological interest can be described by linear laws, as we see in the next section and in the next two chapters. In the final part of the book, we study the implications of including nonlinear terms in the thermodynamic formalism.

6.3 MICHAELIS–MENTEN EQUATION AND COOPERATIVE PROPERTIES OF ENZYMES

A case in which linearity is applicable is the study of the behavior of a system under some very definite conditions. Let us look at an example.

Let us consider the simplest enzyme reaction, which involves conversion of a substrate S into a product P by means of a single process such as

$$F_0 + C \underset{k_1'}{\overset{k_1}{\rightleftharpoons}} F_1$$

$$F_1 \underset{k_2'}{\overset{k_2}{\rightleftharpoons}} F_0 + P$$

where F_0 is the enzyme concentration and F_1 is the concentration of the substrate–enzyme complex. The constants k_i and k_i' represent respectively the rate constants in the forward and backward direction. For the purpose of simplification, let us suppose that $k_2' \approx 0$. Under these conditions, the kinetic equations are

$$\begin{aligned} \dot{C} &= -k_1F_0C + k_1'F_1 \\ \dot{F}_1 &= k_1F_0C - (k_1' + k_2)F_0 \\ \dot{P} &= -\dot{C} = k_2F_1 \end{aligned} \tag{6.13}$$

the total concentration of the enzyme is constant

$$E = F_0 + F_1 = \text{constant}$$

and consequently, in (6.13) the second equation is

$$\dot{F} = k_1EC - (k_1' + k_2 + k_1C)F_1 \tag{6.14}$$

In the stationary state (if $C \gg E$), characterized by

$$\dot{F}_1 = \dot{F}_0 = 0$$

we obtain for F_1

$$F_1 = \frac{k_1EC}{k_1' + k_2 + k_1C}$$

From (6.13), and bearing in mind that $\dot{P} = -\dot{C}$ *coincide with the reaction rate* v, we obtain

$$v = \frac{k_2k_1EC}{k_1' + k_2 + k_1C} \tag{6.15}$$

Dividing the numerator and denominator by k_1 and defining

$$k_M = \frac{k_1' + k_2}{k_1}$$

leads to the Michaelis–Menten equation

$$v = \frac{k_2 CE}{k_M + C} \tag{6.16}$$

This equation is represented graphically in Fig. 6.1, where the maximum velocity $v_{\max}$ can be given by

$$v_{\max} = \lim_{c \to \infty} \frac{k_2 CE}{k_M + C} = k_2 E$$

so that (6.16) can also be written as

$$v = \frac{v_{\max} C}{k_M + C} \tag{6.17}$$

The relation between the reaction rate and the concentration is not linear. But if we study the reaction in zones I or II, linearity is a good approximation, even though it is advisable to remember its many limitations.

In fact, even in very simple processes, conversion of a substrate into a product involves various steps, and the kinetics of enzyme reactions in the steady state often are different from what is represented in Fig. 6.1. Generally, the v vs. C curve displays more complicated behavior, with inflection points, maxima, and so on. In Fig. 6.2, we present a typical example of this situation. In general, this behavior is due to the presence, in the enzyme, of a structure with different interactive subunits such that the behavior of the enzyme is cooperative; that is, the affinity of the substrate and the catalytic activity of one specific subunit depends on the state of the other subunits.

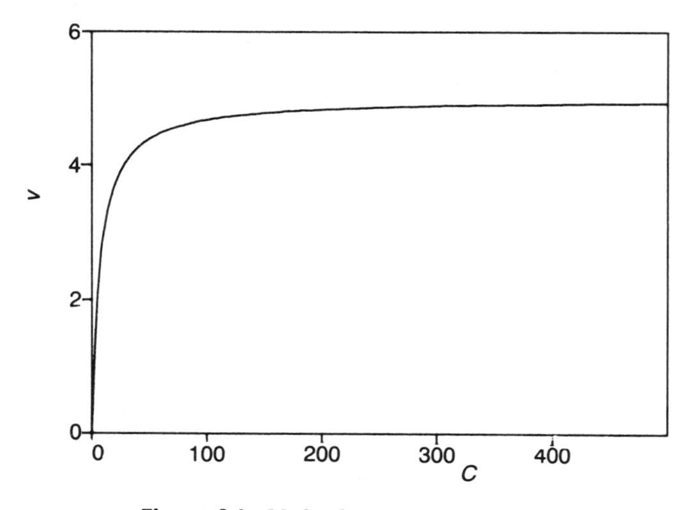

Figure 6.1 Michaelis–Menten equation.

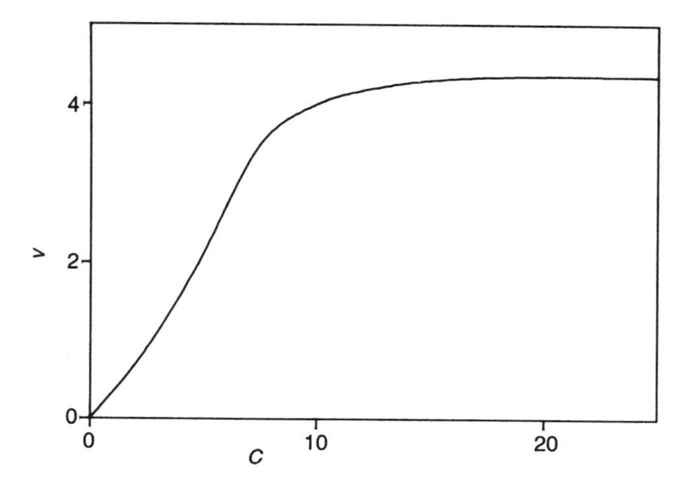

Figure 6.2 The v vs. C curve.

Let us consider a simple model of the enzyme molecule formed by two identical interactive subunits. Each subunit can be occupied by the substrate. The enzyme can be in three states F_{11}, F_{01}, F_{00} corresponding to respectively either two subunits occupied by the substrate, only one occupied by the substrate, or both unoccupied. Under these conditions, we can schematically describe the reaction as the following:

$$\begin{array}{ll}
F_{00} + C \underset{k_1'}{\overset{k_1}{\rightleftharpoons}} F_{01} & F_{00} + C \underset{k_1'}{\overset{k_1}{\rightleftharpoons}} F_{10} \\
F_{01} \xrightarrow{k_2} F_{00} + P & F_{10} \xrightarrow{k_2} F_{00} + P \\
F_{01} + C \underset{k_3'}{\overset{k_3}{\rightleftharpoons}} F_{11} & F_{10} + C \xrightarrow{k_3} F_{11} \\
F_{11} \xrightarrow{k_4} F_{01} + P & F_{11} \xrightarrow{k_4} F_{10} + P
\end{array} \tag{6.18}$$

where for simplification we have assumed that $k_2 = k_4 = 0$; on the other hand, $F_{10} = F_{01}$. The kinetic equations will be

$$\begin{aligned}
\dot{F}_{00} &= -2k_1F_{00}C + 2(k_1' + k_2)F_{10} \\
2\dot{F}_{10} &= 2k_1CF_{00} - (2k_1' + k_2 + k_3C)F_{10} + 2(k_3' + k_4)F_{11} \\
\dot{F}_{11} &= 2k_3CF_{10} - 2(k_3' + k_4)F_{11}
\end{aligned} \tag{6.19}$$

In the stationary state, we have $\dot{F}_{00} = \dot{F}_{10} = \dot{F}_{11} = 0$. Since E, the total enzyme concentration, is constant ($E = F_{00} + 2F_{10} + F_{11} =$ constant), the reaction rate v can be given by

$$v = 2k_2F_{10} + 2k_4F_{11}$$

Starting from (6.19) and the steady state condition, combined with the last equation, we obtain

$$v = 2k_2EC\,\frac{\alpha C + \beta K}{C^2 + 2\beta KC + \beta K^2} \tag{6.20}$$

where the coefficients α, β, K, and K' are defined as

$$\alpha = \frac{k_1}{k_2}, \qquad \beta = \frac{K'}{K}, \qquad K = \frac{k_1' + k_2}{k_1}, \qquad K' = \frac{k_3' + k_4}{k_3}$$

The cooperativity, that is, the interaction between the substrates bound to the enzyme, can be determined by the value of the parameters α and β. If $\alpha = \beta = 1$, Eq. (6.20) coincides with (6.17) since

$$v = 2k_2EC\,\frac{(C + K)}{(C + K)^2} = \frac{2k_2ES}{C + K} = \frac{v_{\max}C}{C + K}$$

that is, follows the behavior described by Fig. 6.1. However, if this condition is not fulfilled, the v vs. C curve has an inflection point, a maximum, or both, depending on the relative values of α and β.

As in the case of the Michaelis–Menten equation, even though the relation between v and C is generally nonlinear, there are zones like the one marked as I in Fig. 6.2, around the inflection point, where we can assume linear behavior. In this case, we can use linear laws with an independent term.

6.4 TWO REACTIONS. DEGREE OF COUPLING

Despite its limitations, the linear theory has appreciable conceptual interest, as we see in the following.

Let us assume two chemical reactions, which we distinguish by the subindices $J = 0, 1$. The two linear constitutive laws are

$$J_0 = L_{00}A_0 + L_{01}A_1 \tag{6.21}$$

$$J_1 = L_{10}A_0 + L_{11}A_1 \tag{6.22}$$

and the corresponding dissipation function $(T\sigma)$ is

$$T\sigma = J_0A_0 + J_1A_1 \geq 0 \tag{6.23}$$

In order for the sum to be positive, we can have either both sums positive, or one of them (A_0J_0) negative while the other (J_1A_1) is positive and large enough to compensate the negative effect of the first term. In the latter case, the first reaction is carried out in the direction opposite to the direction predicted by its affinity—that is, in the direction opposite to what would happen if it occurred spontaneously. This apparently antithermodynamic effect is possible owing to the entropy production due to the other reaction, which certainly occurs in the direction predicted by its affinity.

This coupling is of great interest in biology, since in many situations (synthesis of macromolecules, proteins, active transport, and so on), the synthesis reaction or the transport phenomenon takes place in the direction opposite to that predicted by its affinity. This is possible only because of some metabolic reaction that occurs in the spontaneous thermodynamic direction. Without this coupling process, many biological processes would be impossible.

It is customary to define as the degree of coupling the ratio

$$q_{01} \equiv L_{01}/(L_{00}L_{11})^{1/2} \tag{6.24}$$

The second law imposes $L_{00}L_{11} \geq L_{01}^2$ and therefore the degree of coupling is limited between -1 and $+1$. When $q = +1$, the system is completely coupled and the two processes can be expressed as a single process. When $q = 0$, the two processes are completely uncoupled and do not have any special interest from the standpoint of energy conversion.

6.5 STOICHIOMETRY AND DEGREE OF COUPLING

We call the ratio of the efflux J_0 to the influx J_1 the *stoichiometric ratio*. This ratio indicates how many moles of a certain component must react in reaction 1 in order to produce a certain number of moles of the other component of the forced reaction 0. According to (6.21) and (6.22), we have

$$\frac{J_0}{J_1} = \frac{L_{00}A_0 + L_{01}A_1}{L_{10}A_0 + L_{11}A_1} \tag{6.25}$$

and dividing the numerator and the denominator by $(L_{00}L_{11})^{1/2}$, we obtain

$$\frac{J_0}{J_1} = Z\frac{q + Zx}{1 + qZx} \tag{6.26}$$

where $Z = (L_{00}/L_{11})^{1/2}$ and $x = A_0/A_1$. We observe that when the degree of coupling q goes to zero, $J_0/J_1 = Z^2x$, so we can consider Z as a stoichiometric parameter.

The most important observation in (6.26) is that the stoichiometric ratio depends on the degree of coupling q and on the ratio of the affinities x. The stoichiometric ratio is a parameter that has been very well studied in metabolic processes. How many moles of oxygen must be consumed in oxidation in order for phosphorylation to produce a mole of ATP? How many moles of ATP must be consumed in order to transport a mole of Na

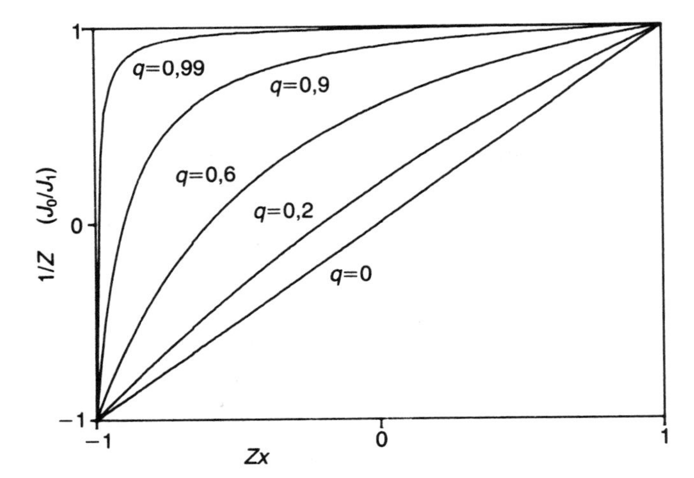

Figure 6.3 Apparent stoichiometry as a function of the affinity ratio and the degree of coupling of two reactions.

against a gradient in active transport? There have always been controversies over the experimental results which affect this question. Equation (6.26) shows that the stoichiometric ratio is a quantity that requires delicate handling, a quantity that involves other phenomena or parameters in addition to the force ratio x. In conclusion, not only should the stoichiometric coefficients or the instantaneous concentrations of the different reactions be considered individually (included in the respective affinities) but also kinetic factors expressed by the parameters L_{00}, $L_{01} = L_{10}$, L_{11} or, more appropriately, by Z and q should be taken into consideration. In Fig. 6.3, we give the stoichiometric ratio for various values of x and q.

6.6 EFFICIENCY AND DEGREE OF COUPLING

Another fundamental quantity or concept in the overall process of energy conversion (oxidative phosphorylation or photosynthesis, active transport, and so on) is the *efficiency*. What fraction of the energy liberated by reaction 1 is used by reaction 0? Unfortunately, the only quantitative expression for efficiency customarily recognized in thermodynamics is the famous Carnot expression studied in Chapter 1. This formula, however, is completely useless in our case, since biological systems are practically isothermal and therefore have nothing to do with the situation described by Carnot. We need other definitions of efficiency that are closer to the specific problem under study.

In the case involving two coupled reactions, one spontaneous and the other forced, it is customary to define efficiency as the rate of free energy consumed in reaction 0 divided by the rate of liberation of free energy in reaction 1, that is,

$$\eta = -\frac{J_0 A_0}{J_1 A_1} \tag{6.27}$$

since $J_1 A_1$ is the energy liberated by reaction 1 and $-J_0 A_0$ is the free energy taken up by reaction 0 per unit time. This ratio is in fact the ratio of the disappearing and evolving energy.

Starting from (6.26), the efficiency η can be given by

$$\eta = -\frac{Zx(q + Zx)}{1 + qZx} \tag{6.28}$$

Now it is a question of knowing for which value of the affinities (that is, the concentrations of the substances that are involved in the reactions) the efficiency of the transfer of energy from reaction 1 to 0 is maximum. To do this, we differentiate η with respect to x; we equate the derivative to zero, and we have

$$x_{\max} = (qZ)^{-1}[-1 + (1 - q^2)^{1/2}] \tag{6.29}$$

The maximum efficiency is obtained by introducing (6.29) into (6.28),

$$\eta_{\max} = q^{-2}[1 - (1 - q^2)^{1/2}]^2$$

whose more common expression is obtained by multiplying the numerator and denominator by $[1 + (1 - q^2)^{1/2}]^2$

$$\eta_{\max} = \frac{q^2}{[1 + (1 - q^2)^{1/2}]^2} \tag{6.30}$$

This ratio tends to zero when the degree of coupling decreases. This is logical since if the reactions are completely uncoupled, the efficiency in the exchange must be zero, as is indeed the case. When the reactions are completely coupled, that is, when $q = +1$, the maximum efficiency is unity—the maximum value that can be achieved (6.27) due to the second law of thermodynamics. This expression for the optimal efficiency has much more practical interest for a biologist than the Carnot formula, but unfortunately is much less well-known.

In many situations, we are interested in knowing what is the efficiency corresponding to the maximum energy production, since this will be the preferred state for the biological system in processes that require great speed, like fleeing and hunting. Then let us maximize $-J_0 A_0$, which is the energy utilized. According to the last expressions, the normalized power is expressed as

$$\frac{J_0 A_0}{L_{11} A_1^2} = -Zx(Zx + q) \tag{6.31}$$

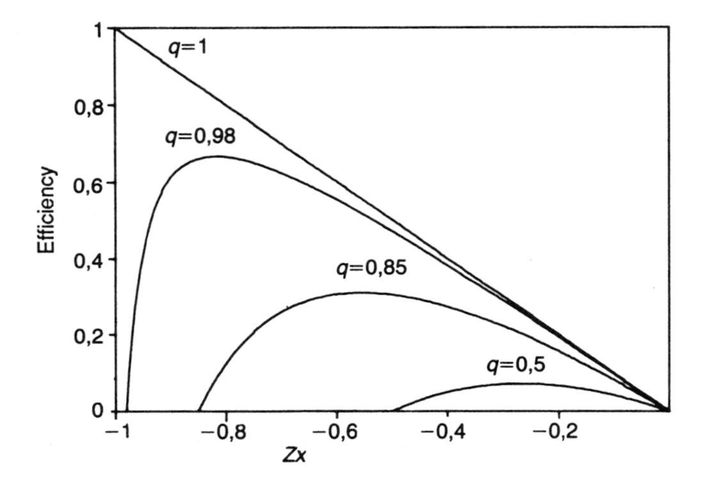

Figure 6.4 Efficiency of the energy conversion of two coupled reactions as a function of the affinity ratio and the degree of coupling.

This energy is maximum for $x' = -q/2z$, and the corresponding efficiency is

$$\eta = (1/4)q^2[1 - (q^2/2)]^{-1} \tag{6.32}$$

In general, then, the efficiency corresponding to the situation of maximum liberated energy will be less than the maximum efficiency, as would be expected. When the coupling is complete, with $q = 1$, we obtain a maximum efficiency of 100 percent, while the efficiency that corresponds to the situation of maximum power is only 50 percent.

Another question of interest is the relationship between the minimum entropy production and the maximum efficiency. The entropy production σ, or better, the dissipation function $T\sigma$, can be written in terms of x and Z as

$$T\sigma = (Z^2x^2 + 2qZx + 1)L_{11}A_1^2 \tag{6.33}$$

For a fixed affinity A_1, this function has its minimum for $x = -q/Z$, which corresponds to $J_0 = 0$. In this state, both the efficiency and the power are zero, according to the definitions we gave in (6.27) and (6.31). The corresponding minimum dissipation is equal to

$$T\sigma = J_1A_1 = (1 - Z^2x^2)L_{11}A_1^2 \tag{6.34}$$

The situation of minimum entropy production does not correspond to the situation of maximum efficiency. Of course, this is also not necessarily the situation of greatest biological or industrial interest.

6.7 CONCLUSIONS

It is important to emphasize how interesting this thermodynamic study of the efficiency of energy conversion is. In conducting a certain experiment, we can interpret the results in terms of the efficiency in the conversion, a viewpoint that will allow us to go into the significance of the results obtained in more depth. We observe that, as in Carnot's result the maximum efficiency depends only on the temperature difference, in this case it depends only on the degree of coupling. This brings out an aspect of the phenomenological coefficients which until now has been little explored. This type of study falls outside the range of the usual studies of chemical kinetics, and finds a more suitable framework in the thermodynamics of irreversible processes.

A second point that must be considered when we speak of efficiency is the fact that living beings do not necessarily have to function in the maximum efficiency zone. Sometimes it is of interest to acquire a large amount of energy in a short time, even if at low efficiency. In an emergency situation, for example, in which it is necessary to flee from some danger, it is easy to understand that survival chooses acquisition of the necessary energy for fleeing, regardless of cost, rather than maximum efficiency. This fact must be kept in mind when we study (and especially when we interpret) efficiencies in biological systems. Another example is provided by situations in which the efficiency is zero, as in the case when $J_0 = 0$ or $A_0 = 0$. These situations often occur in the organism when, for example, it must maintain a certain electrochemical potential difference between the two sides of the membrane, making the net flux of some ions be equal to zero. In other conditions, we are interested in maintaining a certain homogeneity on both sides of the membrane, despite the fact that it is necessary to maintain an energy flow in order to do this. Situations like this can be of vital interest in many cases, even at the sociological scale. For example, it is absolutely necessary to maintain a good university in order for the science of a country to be comparable to its counterparts elsewhere. However, it costs money to maintain this situation of homogeneity at the scientific level, to maintain this university; that is, it absorbs a certain energy flow. Here we have an example in which a situation of possible low efficiency in the short run is vital for the development of the organism (in this case, a social organism) in the long run.

BIBLIOGRAPHY

Caplan, S. R., and A., Essig, *Bioenergetics and Linear Nonequilibrium Thermodynamics. The Steady State.* Cambridge, Massachusetts: Harvard University Press, 1983.

KATCHALSKY, A., and P. F. CURRAN, *Nonequilibrium Thermodynamics in Biophysics*. Cambridge, Massachusetts: Harvard University Press, 1965.

VOLKENSHTEIN, M. V., *Biophysics*. Moscow: Mir, 1983.

WESTERHOFF, N. V., and K. VAN DAM, *Thermodynamics and Control of Biological Free-Energy Transduction*. Amsterdam: Elsevier, 1987.

7

Active Transport

7.1 INTRODUCTION

In the last chapter, we presented some ideas about coupled processes. In this chapter and the next, we study two biological problems in some detail: active transport and oxidative phosphorylation. This will help us develop, apply, and understand the concepts in the last chapter. In Chapters 4 and 5 we studied transport processes, and in Chapter 6 we studied chemical reactions. In this chapter and the following, transport and chemical reactions appear to be intimately coupled. The coupling between a scalar process (like chemical reactions) and a vectorial process (like transport) is not a general phenomenon, but rather is possible only under exceptional circumstances when the system under study has a sufficient degree of anisotropy. In the cases we will study, this connection between a scalar and a vectorial process is provided by biological membranes, with its molecules preferentially ordered in a definite direction perpendicular to the plane of the membrane.

In this chapter, we discuss active transport. Almost all cell membranes, both outer and inner, are endowed with one sort or another of active transport: the sodium and potassium pump in almost all cells, especially nerve cells; the active transport of calcium in muscle cells; the transport of protons in mitochondrial membranes and chloroplasts; the transport of protons in the retina; active transport in various tissues (kidney, stomach, intestinal, and so on). So this is a sufficiently widespread and important phenomenon for us to devote all possible scientific tools, both theoretical and experimental, toward achieving the most complete understanding possible. Here we emphasize some aspects that are of special interest in the thermodynamics of irreversible processes, and we will omit the more biochemical points, which are well-known to students of biology.

7.2 CELLS AS NONEQUILIBRIUM STATIONARY STATES

It is well-known that the chemical composition of the extracellular medium is essentially different from the intracellular medium. The sodium concentration in the exterior is about 145 moles $\cdot$ m^{-3}, while the sodium concentration in the interior is about 12 moles $\cdot$ m^{-3}. The concentration of potassium ions in the exterior is about 4 moles $\cdot$ m^{-3}, while in the interior it is about 155 moles $\cdot$ m^{-3}. If we take the electrical potential outside the cell as the reference ($V_e = 0$ mV), its interior potential is $V_i = -90$ mV. We cite only a few typical data, limiting ourselves to the more essential ones.

We observe that, since the potassium concentration is greater in the interior than in the exterior, there will be an outward diffusion flow of potassium. On the other hand, since the interior potential is less than the exterior potential, and since the K^+ ions are positive, there will be an inward flow of this element (of electrochemical origin). In order to determine which of these two flows is greater, that is, which predominates (the outward diffusion flow or the inward electrochemical flow), we first need to examine the equilibrium situation. This circumstance is achieved when the electrochemical potential of potassium is the same in the interior as in the exterior, that is, according to (2.23), when

$$\begin{aligned} \mu_{\text{int}} &= \mu_0(T) + RT \ln c_i + \mathcal{F} z V_i \\ \mu_{\text{ext}} &= \mu_0(T) + RT \ln c_e + \mathcal{F} z V_e \end{aligned} \tag{7.1}$$

are equal. In (7.1), c_i, c_e are the exterior and interior concentrations respectively, R is the gas constant, T is the absolute temperature, $\mathcal{F}$ is the electric charge per mole or the faraday, z is the number of charges per ion (+1 in the case of potassium), and the V values are the respective electrical potentials, while $\mu_0(T)$ depends only on temperature (and not on concentration or voltage). Thus, in equilibrium we obtain

$$RT(\ln c_i - \ln c_e) = \mathcal{F} z (V_e - V_i)$$

that is, the potential difference in equilibrium must be

$$V_e - V_i = (RT/\mathcal{F} z) \ln(c_i/c_e) \tag{7.2}$$

This equation, called Nernst's law, defines the potential difference necessary for the inward electrochemical flow to counteract the outward diffusion flow, which is determined by the concentrations c_i and c_e. This law can also be written, if we divide the numerator and denominator of the right-hand side by Avogadro's number, as

$$V_e - V_i = (kT/q) \ln(c_i/c_e) \tag{7.3}$$

where k is the Boltzmann constant ($k = 8.62 \cdot 10^{-5}$ eV/K) and q is the charge of each ion of the species under consideration. In the case of potassium, $q = +e$, with e equal to the charge of the electron in absolute units.

Which flow predominates in the case of potassium in the last situation? First let us calculate the equilibrium potential according to the Nernst equation. We have

$$V_e - V_i = (8.62 \times 10^{-5}\ \text{eV} \cdot \text{K}^{-1} \times 310\ \text{K} \times 1\ e^{-1} \ln(155/4) = +98\ \text{mV}$$

We have assumed that $T = 310$ K, that is, that the temperature is about 37°C, the usual temperature of the human organism. In order for the electrochemical flow to exactly compensate the exit of ions due to diffusion, it is necessary that the interior potential be about $V_i = -98$ mV (since $V_e = 0$ mV, by agreement). However, the true interior potential is only $V_{i\text{true}} = -90$ mV; that is, it is not negative enough to bring in the same amount of potassium as it leaves, resulting in a higher outward flow and, in short, in the tendency of potassium to leave the cell.

For sodium, the situation is more clear. The diffusion flow is inward, since the exterior concentration is greater than the interior concentration. On the other hand, the flow of electrochemical origin is also inward, since Na^+ is positive and tends to move toward the zones of lower potential. Thus the net flow of sodium is inward. In principle, the tendency of sodium to enter the cell is much greater than the tendency of potassium to leave the cell, since in the case of sodium the two driving forces for entering are added together, while in the case of potassium one is subtracted from the other. Nevertheless, we observe that the number of Na^+ ions that enter is on the same order of magnitude as the number of K^+ ions that leave. This leads to the conclusion that the permeability of the cell membrane to sodium ions is much less than the permeability to potassium.

The cell, then, is not found in an equilibrium state, and not only with respect to the aspects which we have considered. Sodium tends to enter, potassium tends to leave; and, if no pump operates, at the end of a certain time the sodium and potassium concentrations would reach some equilibrium values, which are far from the observed values. Therefore, with respect to sodium and potassium, the cell at rest is found in a nonequilibrium stationary state. This state is maintained by the sodium–potassium pump, which pumps out the entering sodium ion and pumps the leaking potassium back into the cell interior, at the expense of a certain metabolic consumption.

This simple introduction provides us with an example of the interest that nonequilibrium stationary states have for the study of biological systems, which, as we see, can in no way be considered as equilibrium systems. The balance of sodium and potassium concentrations (a simple part of a more complex system) is only one of many examples of such nonequilibrium stationary states in biology.

7.3 ELECTRICAL MODEL OF ACTIVE TRANSPORT

Here we limit ourselves to considering the case in which only one chemical reaction is responsible for active transport of a single chemical species. Many pumps for active transport in an organism are more complicated, but in this introductory text we do not consider them. Since the cell is maintained in a nonequilibrium stationary state, as a result of the sodium–potassium pump, the quantitative characterization of this mechanism is very important for the description of the resting state of the cell. Let us begin with a very simple model, a simple analogy to an electrical circuit, which will help us understand the situation. Then we will see the advantages provided by the description in terms of the thermodynamics of irreversible processes.

The electrical circuit to which we refer is represented in Fig. 7.1. This model is described by three parameters: R_p is the passive resistance to the inward passage of the Na^+ ions, that is, is related to the permeability of the membrane to sodium; R_a is the resistance to the passage of sodium by the channels through which these ions are pumped toward the outside—or, in other words, the internal resistance of the sodium pump. Finally, E_{Na} is the electromotive force that describes the sodium pump; in this case, this is interpreted in analogy to an electrical battery that moves ions from the interior to the exterior. In an electric battery or cell, a chemical reaction creates the electromotive force that moves the electrons in the electrical circuit. In the sodium pump, a metabolic chemical reaction provides the energy, or the electromotive force, for expelling the sodium.

Here we will not study which specific reaction is involved, since this is a problem sufficiently well studied in biochemistry. The role of bio-

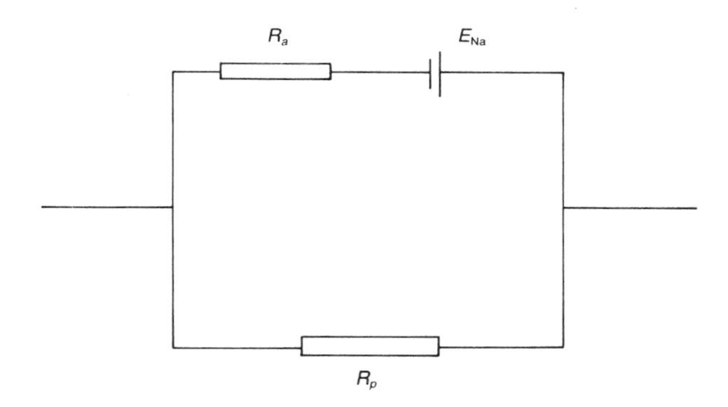

Figure 7.1 Electrical circuit analogous to the process of active transport. E_{Na}, electromotive force of the sodium pump; R_a, internal resistance of the pump; R_p, resistance of the passive channels.

physics at this point is to bring out the essential outlines of the biological mechanisms. A next step, once we have achieved clear understanding of the mechanism, is to concern ourselves about which specific materials realize the mechanism and how to carry it out. This last point goes beyond the level that we have set for this text.

Now let us see how the electrical analogy provides a model that allows us to include and systematize the experimental data and make comparisons between the measurements of various authors. The basic problem of this model is that the electromotive force, which in principle should be a purely energetic parameter, also includes permeability factors. Ideally, on the other hand, there should be a clear distinction between energetic phenomena and the kinetic factors, which should be included exclusively in the resistances R_a and R_p. This confluence of various factors in a single parameter, E_{Na}, is brought out when, for example, we add aldosterone to the system. In this case, the affinity of the chemical reaction of the "cell" increases, but its electromotive force does not change. This fact is explained by supposing that the aldosterone, despite increasing the affinity of the reaction (a fact which in principle should increase the electromotive force of the hypothetical cell), decreases the permeability of the membrane. The two factors counteract each other, then, in the electromotive force. It is also possible to give a theoretical proof that energetic (chemical) and kinetic (permeability) factors are included in the parameter E_{Na}, but the proof is long and we omit it for the sake of simplicity.

Finally, the electrical model, despite its ease of visualization, does not bring out with sufficient clarity the degree of coupling between the chemical reaction and the transport. This is obvious, since we know only up to what point E_{Na} contains energetic and kinetic factors. As we have seen in the last chapter, the degree of coupling plays an important role when we try to determine the efficiency and the apparent stoichiometry of the combined process. These deficiencies are avoided and corrected in the formalism of the thermodynamics of irreversible processes.

7.4 THERMODYNAMIC MODEL OF THE SODIUM PUMP

The thermodynamic model of active transport in general, and of the sodium pump in the particular case we have discussed, must involve two processes: a chemical reaction described in the thermodynamics of irreversible processes by the thermodynamic force or the corresponding affinity (which, we recall, depends on the concentrations of the species involved and the temperature), and the sodium flux which must be associated with, as the thermodynamic force, the electrochemical potential difference for sodium on both sides of the membrane. That is,

$$X_{Na} = \mu_{Na\ ext} - \mu_{Na\ int}$$

where the electrochemical potentials have the form given in (7.1).

The corresponding thermodynamic fluxes are J_r, which indicates the rhythm of the metabolic reaction or the number of moles transformed per unit time (the number of moles of ATP consumed per second, for example), and J_{Na}, or the sodium flux that crosses the membrane per unit time.

The phenomenological laws are, in this case, in accord with the formalism presented in Chapter 3,

$$J_r = L_{Na_r} X_{Na} + L_r A \quad (7.4)$$

$$J_{Na} = L_{Na} X_{Na} + L_{Na_r} A \quad (7.5)$$

Note that, due to the Onsager relations, the four phenomenological coefficients which in principle we should include are reduced to three: L_{Na} is the passive permeability to sodium; L_r determines the rhythm at which the metabolic reaction occurs if there is no sodium transport; L_{Na_r} is the coupling coefficient between the chemical reaction and the sodium flux. The second term in Eq. (7.5) describes the flux of sodium transported actively as a result of the chemical reaction of the pump, while the first term on the right-hand side indicates the passive flux of sodium toward the cell interior. In Table 7.1, we give the values of the phenomenological coefficients for a certain membrane.

We can prove that the relation between the parameter E_{Na} (or the electromotive force in the electrical model) and the affinity A in the thermodynamic model can be given by

$$E_{Na} = (1/\mathcal{F})\,(L_{Na_r}/L_{Na})A \quad (7.6)$$

where we clearly see in this case the confluence in E_{Na} of the energy factors described by A (the affinity of the chemical reaction) and the kinetic factors (the permeability and the degree of coupling). In the thermodynamic model, these various factors are well differentiated. In ad-

TABLE 7.1 Phenomenological Coefficients of the Sodium Pump in the Urinary Bladder Membrane of the Toad

Coefficient	Value	Units
L_{Na} =	103.9 + 12.5	$mole^2 \cdot cm^{-2} \cdot s^{-1} \cdot kcal^{-1}$
L_{Na_r} =	5.41 + 0.33	$mole^2 \cdot cm^{-2} \cdot s^{-1} \cdot kcal^{-1}$
L_r =	0.369 + 0.073	$mole^2 \cdot cm^{-2} \cdot s^{-1} \cdot kcal^{-1}$

dition, the degree of coupling is expressed directly through the coefficient L_{Na_r}. In this case, the degree of coupling, as we have defined it in the preceding chapter, is

$$q = L_{Na_r}/(L_r L_{Na})^{1/2} \tag{7.7}$$

In the example presented in Table 7.1, the degree of coupling is $q = 0.86$, and the maximum efficiency in this situation is, according to (6.21), about 33 percent.

7.5 COMMENTS ON STOICHIOMETRY AND EFFICIENCY IN ACTIVE TRANSPORT

Many authors have concerned themselves with the stoichiometry of active transport. How many moles of ATP are consumed per mole of sodium transported actively? The answer found in treatises on biochemistry is "one-third." However, this answer is incomplete. In reality, the number of moles of ATP that must be consumed per mole of sodium depends on the rate at which the sodium is transported. We already noted this dependence of the stoichiometry on the reaction rate in the last chapter. Here we try to give a more precise idea about the situation.

If a combined process is a chemical reaction, then certainly it will have a fixed stoichiometry. This will be the case if the degree of coupling is unity. The coupling between these two processes (reaction and transport) is not purely chemical, but rather energetic. The reaction supplies a certain amount of energy, part of which passes to the sodium which, using this energy, can be transported against a gradient—that is, in an antispontaneous direction. The fact that the coupling is energetic explains why there are so many precautions, and so many discrepancies between results, when we deal with apparent stoichiometry. Perhaps in some cases the fraction of the energy liberated by the first reaction and used by the transport is different from the fraction used in other circumstances. For this reason, the apparent stoichiometry depends on the degree of coupling.

On the other hand, the hydrodynamic resistance force that an ion experiences when crossing the membrane is proportional to its velocity, according to (4.8). This means that the energy with which the ion must be provided in order to cross the membrane at different velocities will also be different. Therefore, in principle it requires more energy to transport one mole per second than one mole per minute, since the velocity of the ions in the first case is higher than in the second case. Thus the metabolic consumption can depend, in principle, on the rate at which transport is accomplished.

Equations (7.4) and (7.5) allow us to evaluate the number of moles and the free energy consumed in the stationary state. In this state, the inward sodium flux ($L_{Na}X_{Na}$) is compensated by the outward flux driven by active transport ($L_{Na,r}A$) such that

$$L_{Na}X_{Na} + L_{Na,r}A = J_{Na} = 0 \tag{7.8}$$

When speaking of efficiency in the last chapter, we referred to the fact that in some processes we are not interested in maximum energy efficiency, but rather in maintaining a certain imbalance in a steady-state situation. Here, the imbalance is given by the value of X_{Na}, or the electrochemical potential difference for sodium between the two sides of the membrane. According to the purely energetic definition (6.8) of the efficiency, this will be zero in the case when $J_{Na} = 0$. It is easy to understand that in this case we are not interested so much in an energetic definition of efficiency, but rather in a new definition like the flux efficiency

$$\eta_{flux} = \frac{\text{active flux}}{\text{free energy consumed}} = \frac{L_{Na,r}A}{J_rA} \tag{7.9}$$

In this definition, we divide what we are interested in obtaining (the flux of active transport) by what we consume in order to obtain it (the metabolic energy). In this stationary state, the efficiency will then be

$$\eta_{flux} = \frac{L_{Na,r}}{L_rA + L_{Na,r}X_{Na}} = \frac{Z_q}{(1 - q^2)A} \tag{7.10}$$

where q is defined by (7.7) and $Z = (L_{Na}/L_r)^{1/2}$.

The metabolic energy consumed per unit time in the stationary state will be, finally,

$$J_rA = (L_rA + L_{Na,r}X_{Na})A$$

and then, according to (7.8), $A_{st} = -(L_{Na}/L_{Na,r})X_{Na}$, and

$$J_rA_{st} = X_{Na}^2L_{Na}(1 - q^2)q^{-2} \tag{7.11}$$

In this relation, we see how, for a given degree of coupling, more energy is consumed when the permeability (L_{Na}) is greater. This is logical, since as more sodium enters, more sodium must be expelled. In addition, the greater the electrochemical potential difference (X_{Na}), the more metabolic energy is consumed in this case, since on the one hand the tendency of sodium to enter increases (since the sodium which enters per unit time is proportional to X_{Na}); and on the other hand it is more difficult to expel the sodium, since the energy necessary to expel it is also related to X_{Na}.

7.6 FINAL COMMENTS: STABILITY

The resting state is not always stable. Nerve cells can transmit electrical impulses because of this fact. The action potential that constitutes the neural signal is initiated when a perturbation of the stationary state exceeds a certain threshold value or minimum level. Then the sodium channels abruptly open collectively, and the sodium penetrates en masse into the cell. The cell is depolarized, that is, its interior electrical potential, which at rest is about -90 mV, changes sign, reaching about 40 mV. At this moment, the potassium channels open and the potassium, finding a positive potential in the interior, leaves the cell. Finally, the sodium and potassium channels close and the sodium–potassium pump reestablishes the resting situation described before—that is, it repolarizes the cell. This process can be described quantitatively by means of the Hodgkin–Huxley equations (1952), whose authors received the Nobel Prize in Medicine for this work.

With this commentary, we want to warn the reader that the stationary state is a basic reference point but is not the only theme to study in thermodynamics. Often it is of great interest to consider the alterations of the system when it is removed from the stationary state. This is exactly what allows us to study the response to external stimuli. The stability of the stationary state is then a basic problem in the thermodynamics of living beings: Will a perturbation be slowly cancelled out by the system? Will it lead to drastic changes in the system? In the case of the neural cell membrane, small perturbations are cancelled out. On the other hand, perturbations that exceed a certain threshold are amplified and constitute the action potential of the neural signal. In the last part of this book we will discuss, as simply as possible, some topics in nonequilibrium stability.

BIBLIOGRAPHY

Caplan, S. R., and A. Essig, *Bioenergetics and Linear Nonequilibrium Thermodynamics. The Steady State.* Cambridge, Massachusetts: Harvard University Press, 1983.

Mikulecky, D. C., "Biological aspects of transport." In *Transport Phenomena in Fluids* (H. J. M. Hanley, ed.). New York: Dekker, 1969.

8

Oxidative Phosphorylation

8.1 INTRODUCTION

In the preceding chapter, we studied the active transport of ions by means of certain "pumps" that consume metabolic energy. In this situation, we had two coupled processes: a chemical reaction and a transport flow. In this chapter, we study a theoretical model of the irreversible processes of oxidative phosphorylation. In this case, we have two reactions (oxidation and phosphorylation) and a transport process (proton transport). All together, then, we have three coupled processes. In many cases, we should consider the transport of other ions, like potassium, but here we limit ourselves to the most essential and irreducible characteristics of the overall process.

The process of phosphorylation (conversion of ADP to ATP) is basic in storage of metabolic energy. ATP, in reconverting to ADP, returns the stored energy and supplies it to various points of the organism in order for the organism to carry out vital functions. In order for ATP to be in a condition to liberate energy, it is essential that it not be in equilibrium with ADP, since if it is in equilibrium with ADP then it will not be hydrolyzed and consequently will not liberate energy (about 30.7 kJ per mole of ATP that is hydrolyzed).

Where does the energy stored in ATP come from? The most efficient process for the formation of this compound is carried out in the mitochondria of the cells, starting from oxidation of various materials. The energy given up by the electrons in the oxidation–reduction process is transmitted in some way, with surprisingly high efficiency, until it is stored in the ATP. This type of phosphorylation is called oxidative phosphorylation. There is another very similar type of phosphorylation, photosynthetic phosphorylation, which occurs in the chloroplasts of plant cells. In this case, the energy comes from light instead of from oxidation. Because of the enormous importance of this process, much research effort has been dedicated to it. Any treatise on biochemistry or bioenergetics gives sufficiently detailed explanations of the process.

The problem that concerns us here is above all physical, and we can construct an abstraction (in the first stage) of the specific type of chemical species involved; our purpose is to develop a scheme for the general mechanism of this process. Here we limit ourselves to an elementary treatment intended to be instructive. The study of photosynthetic phosphorylation is analogous in principle, but with the difficulty that the thermodynamic treatment of electromagnetic radiation is very subtle, and a rigorous theory is available only in the case when the radiation has spectral characteristics corresponding to a black body (which is not the case at all for the light received by chloroplasts). This difficulty is not completely hopeless, but it entails some subtleties with which we have no intention of getting involved. The broad outlines, however, are similar in these two types of phosphorylation.

Two reactions occur: oxidation and phosphorylation. Of course, the big problem is to see how the two processes are coupled. There are various types of hypotheses concerning this point: *the chemical hypothesis*, which postulates creation of a high-energy intermediate compound; and *the chemiosmotic hypothesis* proposed by Mitchell in 1961 (Nobel Prize in Chemistry in 1978 for this work), according to which the coupling is due exclusively to a flow of protons which, driven by oxidation, pass from the interior to the exterior of the inner membrane of the mitochondrion and which, going back to the interior, in turn drive the phosphorylation process. Both the chemical theory and the chemiosmotic theory assume this flow of protons. The difference between the two theories involves the fact that, in the chemical theory, this flow is considered accessory (due to hydrolysis of the high-energy intermediate) while in the chemiosmotic theory this flow is essential, since it is the only link between oxidation and phosphorylation. In favor of the chemiosmotic hypothesis is the fact that no high-energy intermediate has been discovered, as well as the enormous importance of the inner mitochondrial membrane in this process, a fact which is used more directly by the chemiosmotic hypothesis. Finally, let us mention *the conformational theory*, in between the other two cited, according to which the intermediate can be an excited state of some molecule associated in some way with the shuttling of protons. This can be the case, for example, in the transport of protons by means of solitary waves in alpha helices, a transport that is carried out with very little dissipation of energy.

As we have already commented, the central questions in the study of thermodynamics are those related to the efficiency of the energy transformation. It is possible to evaluate this efficiency starting from the free energies of the reactions involved. Thus, one of the typical oxidation reactions is

$$\mathrm{NADH} + \mathrm{H}^{+} + \tfrac{1}{2}\mathrm{O}_2 \longrightarrow \mathrm{NAD}^{+} + \mathrm{H_2O} + 221.3\ \mathrm{kJ}$$

while phosphorylation can be described as

$$3\mathrm{ADP} + 3\mathrm{Pi} \longrightarrow 3\mathrm{ATP} + 3\mathrm{H_2O} - 92\ \mathrm{kJ}$$

The efficiency of the process is calculated as the ratio 92 kJ/221.3 kJ, about 41 percent. The rest of the energy is dissipated as heat. This efficiency has been calculated by means of equilibrium thermodynamics. Nevertheless, the real cellular situation is far from an equilibrium situation, so that in general the efficiency depends on the rate at which these processes occur.

Another characteristic of interest, related to the efficiency, is the stoichiometry. In the indicated reactions, the ratio P/O is 3.0; in other words, on the average one mole of O_2 is consumed in order to produce three moles of ATP. However, this coefficient is not fixed. It would be fixed if both reactions were chemically coupled, but they are energetically coupled. This type of coupling, as we have said, does not allow for a fixed stoichiometry, since the fraction of energy used by the second reaction can depend on very different circumstances.

8.2 ELECTRICAL MODEL FOR OXIDATIVE PHOSPHORYLATION

As in the last chapter, let us begin with formulation of a model analogous to an electrical circuit. Despite the limitations already mentioned in the case of active transport, this model allows us to simply visualize some ideas and serves as a point of comparison with the thermodynamic description of the last section.

The parallelism between the electrical circuit and oxidative phosphorylation (Figs. 8.1 and 8.2) is the following: The battery V_0 drives the passage of current through the circuit, transferring energy to the charges. At the same time, the battery V_p opposes the passage of current, that is, pushes the charges in the opposite direction, expending some of the energy of the charges that pass through it. Both batteries have a certain internal resistance R_0 and R_p respectively. Finally, some of the current flow in the circuit can pass through the passive resistance of the membrane R_m due to the pores of the membrane (inversely proportional to the permeability, as in passage of protons), in which case it does not transfer energy to battery, V_p, but rather converts it into heat in R_m by the Joule effect.

The situation is analogous in phosphorylation: the oxidation reaction expels protons toward the exterior of the inner mitochondrial membrane. Three centers of the pump have chains of electrons; each center expels two protons for each electron that crosses it. Since the exterior of the membrane is at a higher potential than the interior (about 200 mV, approximately), this process transfers energy to the protons. These pro-

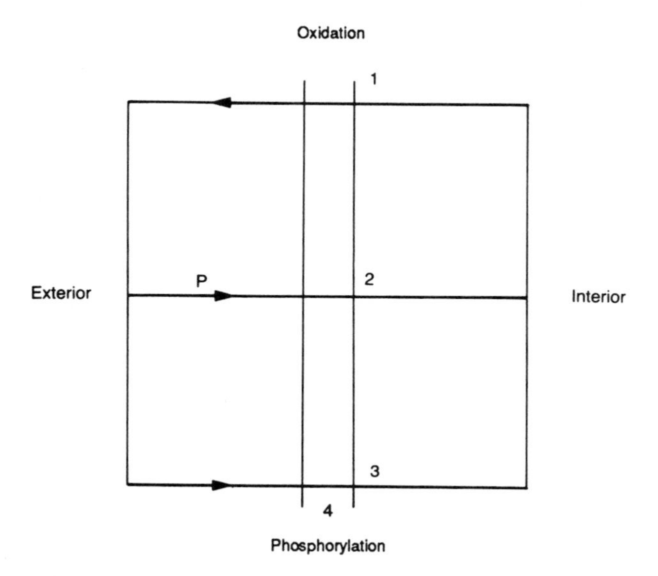

Figure 8.1 Chemiosmotic model of oxidative phosphorylation. (1) Chain of electron transport; (2) passive pore of the membrane; (3) ATPase; (4) inner mitochondrial membrane; p, — proton flux.

tons can return to the interior, passing through thhe ATPase channel, with which a phosphorylation is produced for every two protons. In this way, the protons give up some of their energy to the ATPase in order to do work. It is also possible that some protons cross the membrane through passive pores, losing their energy as heat.

This analogy clarifies the following points: On the one hand, it emphasizes the importance of the membrane, since an injury to the membrane would open up pathways for protons through which they could circulate without phosphorylating. This phenomenon is well-known in practice, and was one of the first pieces of material evidence for the chemiosmotic theory. Second, there are substances (like dinitrophenol) that

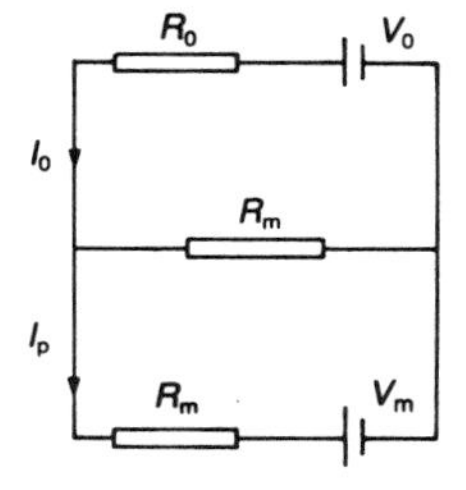

Figure 8.2 Electrical circuit for oxidative phosphorylation. V_0: oxidation potential; V_p: phosphorylation potential; R_0: resistance of the oxidation pump; R_p: resistance from ATPase; R_m: passive resistance of the membrane to the proton flux.

can become encrusted on the membrane, creating channels (or translocation pathways) for the protons. As is logical, this reduces the efficiency and the stoichiometry of the conversion in such a way that less energy is stored in the ATP but more heat is produced. Some tissues whose role is specifically thermogenic, like brown adipose tissue, especially exercise this function. On the other hand, some animals increase the permeability of the mitochondrial membrane at the time of hibernation. As a result, oxidation produces heat to maintain certain minimum temperatures but produces hardly any ATP, since under hibernation conditions (with activity reduced to a minimum) the metabolic demands are very small. When Spring arrives, the permeability of the membrane decreases and the situation reverses itself, now making available sufficient ATP to the animal for its daily activities.

The protons do not always circulate in the direction that we have just indicated. In the submitochondrial particles and in the chloroplasts, they circulate in the opposite direction; that is, the photochemical reaction makes the protons penetrate the inner membrane and, when they leave by means of the corresponding ATPases, they drive the phosphorylation.

The treatment of this elementary circuit by means of Kirchhoff's laws leads to the following equations for I_0 (the current of protons expelled through the electron chain) and for I_p (current of protons which cross by the ATPase):

$$I_0 = \Delta^{-1}(R_p + R_m)V_0 + \Delta^{-1}R_m V_p \tag{8.1}$$

$$I_p = \Delta^{-1} R_m V_o + \Delta^{-1}(R_0 + R_m)V_p \tag{8.2}$$

where $\Delta = R_o R_p + R_o R_m + R_m R_p$. If n_o and n_p are the stoichiometries of the respective pumps ($n_o = 6$ protons/electron, $n_p = 1$ ATP/2 protons), we can relate I_o and I_p to the number of electrons that flow through the electron chain and to the number of ATP molecules formed in the ATPases per unit time and per unit area, if R_o, R_m, and R_p are defined per unit area. Typical values for these parameters are $R_0 = 1250\ \Omega$, $R_p = 660\ \Omega$, $R_m = 2500\ \Omega$ per milligram of membrane protein and $V_p \approx 0.2$ V and $V_o \approx 0.8$ V.

Finally, we observe the symmetry relation in the coefficients which in (8.1) and (8.2) relate the fluxes I_o and I_p with the forces V_o and V_p. The cross coefficient, in fact, is $\Delta^{-1}R_m$ in both equations.

The electrical model has conveniently provided us with exactly the level of simplicity in the description that we want. It does not give biochemical details for the various reactions or the structural configurations of the molecules, but simply the energy balance, the stoichiometry, and the efficiency of the energy conversion.

8.3 THERMODYNAMIC MODEL FOR OXIDATIVE PHOSPHORYLATION

Obviously, as we have just discussed, it is necessary to consider at least three processes in oxidative phosphorylation: oxidation, transport of protons, and phosphorylation. Let us associate a flux with each of these: J_0, J_H, and J_p respectively, which indicate the number of moles transformed or transported per unit time and per unit mass of mitochondrial protein (as is more frequently done). There are also three independent thermodynamic forces: the affinities of the reactions of oxidation A_0 and of phosphorylation A_p, and the electrochemical potential difference for the protons $\Delta\mu_H$ between the exterior and the interior of the mitochondrial membrane. This difference can be expressed in terms of the potential difference ΔV and the concentration difference, or the pH difference, as

$$\Delta\mu_H = \mathcal{F}\Delta V - 2.3RT\Delta \text{pH} \qquad (8.3)$$

where $\mathcal{F}$ is the faraday (96,500 coulombs $\cdot$ mole^{-1}); the factor 2.3 is due to use of the decimal logarithm instead of the natural logarithm.

As we have seen in the general theory, the linear thermodynamics of irreversible processes proposes the following relations between the thermodynamic fluxes and forces:

$$J_0 = L_{00}A_0 + L_{0H}\Delta\mu_H + L_{0p}A_p \qquad (8.4)$$

$$J_H = L_{H0}A_0 + L_{HH}\Delta\mu_H + L_{Hp}A_p \qquad (8.5)$$

$$J_p = L_{p0}A_0 + L_{pH}\Delta\mu_H + L_{pp}A_p \qquad (8.6)$$

where the phenomenological coefficients L_{ij}, which we assume satisfy the Onsager reciprocity relations, contain the general information on the coupling mechanisms. We observe that, according to the Onsager relations, we have only six independent coefficients out of the nine that appear in (8.4)–(8.6).

These coefficients are subject to various restrictions. On the one hand, the second law of thermodynamics imposes the requirement that L_{00}, L_{HH}, and L_{pp} be positive, and that $L_{00}L_{HH} \geqslant L_{0H}^2$ and $L_{pp}L_{HH} \geqslant L_{pH}^2$. The coefficient L_{0H} couples oxidation with proton transport; at the same time, the equality $L_{0H} = L_{H0}$ implies reversibility of this mechanism, that is, that a proton flux which passes in the opposite direction, from the exterior to the interior, by means of the pumps in the electron chain, causes reversal of the chemical reaction of oxidation. This idea of reversibility is also displayed in the equality between the coefficients $L_{Hp} = L_{pH}$, the first of which indicates how many protons are displaced toward the outside if hydrolysis of ATP occurs instead of phosphorylation, and the second of which indicates how many moles of ATP are produced when the protons enter the inner mitochondrial membrane. Finally, the coefficients $L_{Op} = L_{pO}$ represent direct coupling between oxidation and phosphorylation. According to the sign criteria that we have adopted, we observe that $L_{OH} \leqslant 0$, $L_{Op} \geqslant 0$, and $L_{pH} \geqslant 0$. In fact, if the oxidation reaction progresses in the positive direction, the proton flux should be negative (toward the exterior). On the other hand, a positive progress for phosphorylation is linked with a positive proton flux (toward the interior), so that the coupling coefficient should be positive. Finally, the progress of the oxidation reaction is linked with the progress of phosphorylation in such a way that $L_{Op} \geqslant 0$.

In principle, both the chemical hypothesis and the chemiosmotic hypothesis can be described by means of these equations. The difference between the two rests on the fact that the second, more radical hypothesis states that $L_{Op} = L_{pO} = 0$, since according to this theory there is no direct coupling between oxidation and phosphorylation. In this way, the only coupling that exists between them is accomplished by means of the proton flux, that is, the coefficients L_{OH} and L_{pH}. Thus, if the chemical potential of the protons is the same in the exterior as in the interior, such that $\Delta\mu_H = 0$, oxidation and phosphorylation will be uncoupled according to the chemiosmotic theory but not according to the chemical theory.

In order to understand in more detail the significance of the coefficients, let us write the net proton flux J_H as the sum of the flux due to oxidation (negative, since it is directed to the left) plus the fluxes (positive) through ATPase and the passive channels. Then we have

$$J_H = -n_O J_O + C_H \Delta\mu_H + n_p J_p \qquad (8.7)$$

where n_O and n_p are the stoichiometric coefficients of the respective pumps, already defined, and C_H is the permeability of the membrane per unit area. Introducing expressions (8.4) and (8.6) for J_O and J_p into (8.7) and comparing the result with expression (8.5) for J_H, and supposing in addition (for simplification) that $L_{Op} = L_{pO} = 0$ in accord with the chemiosmotic hypothesis, we can make the identification $L_{OH} = -n_O L_{OO}$, $L_{pH} = n_p L_{pp}$, and $L_{HH} = C_H + n_p^2 L_{pp} + n_O^2 L_{OO}$. It is obvious that these coefficients satisfy the thermodynamic restrictions already mentioned. The system is thus described by five parameters: the two stoichiometric coefficients (n_O and n_p) and the three coefficients L_{OO}, L_{pp}, and C_H. The sixth coefficient, L_{Op}, is zero in this case. In Table 8.1, we give the values of L_O, L_p, and C_H in two tissues.

Recall that "nanoatom" or "nmole" refers to the nanogram–atom or the nanomole (nano $= 10^{-9}$), and that the mg in the denominator stands for mg of mitochondrial protein. Observe how, in the second tissue (which has a thermogenic function), the permeability C_H is much greater than in the first tissue (which has an energetic function).

TABLE 8.1 Values of the Phenomenological Coefficients ($mg^{-1} \cdot min^{-1}$ mV^{-1})

Coefficient	Mitochondria of rat liver	Mitochondria of hamster brown adipose tissue
L_0	1.9 nanoatoms of O_2	0.5 nanoatom O_2
L_p	7.9 nmole ATP	0.4 nmole ATP
C_H	3.2 nmole H^+ ATP	35 nmole H^+ ATP

Two problems that arise are the validity of the linear relations on the one hand, and of the reciprocity relations on the other hand. As we have already commented, the phenomenological equations of the chemical reactions are often not linear, and we can expect them to be linear only when $A \ll RT$, that is, when the affinity is much less than the thermal energy. The biological situation is very complex, with biochemical reactions with high affinities, and nevertheless the experimental results in the case of oxidative phosphorylation are compatible in many cases with a linear law, even though the affinity of the process in reality is much greater than RT. There are three types of possible explanations: (1) the flux–force relation in enzymes is often of a sigmoidal type, with an inflection point; when linearizing the relation at this inflection point, we set the second derivative (that is, the second-order terms) to zero, as we have said in Chapter 6. (2) Other researchers have proposed that this is due to the fact that, even though the overall affinity A is much greater than RT, the overall reaction consists of a series of steps, at least one of which has low affinity ($A \ll RT$), so that, as the limiting step, it determines the linear character of the combined law. (3) A third hypothesis, more exciting than the previous ones but more limited and speculative, suggests that the linearity of the thermodynamic equations of oxidative phosphorylation is due to a complicated process of efficiency optimization. This hypothesis is based on calculations which prove that for a sufficiently general type of biological energy converter far from equilibrium, the efficiency is optimized when the processes are linear. This makes it of interest to attribute a special significance (somewhat teleological) to the observed linearity.

8.4 STOICHIOMETRY OF ENERGY CONVERSION

The equations of the thermodynamics of irreversible processes allow us to study the overall stoichiometry and the efficiency of various partially coupled reactions, in the case when the coupling is not directly chemical but rather energetic. Here we will study the stoichiometry within the chemiosmotic hypothesis, as an application of the previous equations. For greatest clarity, we will next rewrite the previous equations (8.4)–(8.6), bearing in mind the identification of coefficients made in the last section:

$$J_O = L_O A_O - n_O L_O \Delta\mu_H \tag{8.8}$$

$$J_H = -n_O L_O A_O + (C_H + n_O^2 L_O + n_p^2 L_p)\,\Delta\mu_H + n_p L_p A_p \tag{8.9}$$

$$J_p = n_p L_p \Delta\mu_H + L_p A_p \tag{8.10}$$

where, for greater convenience, we have written $L_O \equiv L_{OO}$ and $L_p \equiv L_{pp}$.

In the steady-state situation, when all the protons that leave reenter, the net flux J_H is zero, so Eq. (8.9) allows us to write $\Delta\mu_H$ as a function of A_O and A_p. Then we obtain, introducing this result into (8.8) and (8.10),

$$J_O = (L_O - L_O^2 n_O^2 L_H^{-1})A_O - n_O n_p L_O L_p L_H^{-1} A_p \tag{8.11}$$

$$J_p = -n_O n_p L_O L_p L_H^{-1} A_O + (L_p - n_p^2 L_p^2 L_H^{-1})A_p \tag{8.12}$$

The overall stoichiometry of the process, that is, the ratio ATP/O_2, can be given bearing in mind the expression

$$L_H = C_H + n_O^2 L_O + n_p^2$$

in Eqs. (8.11) and (8.12), by

$$\frac{ATP}{O_2} = \frac{J_p}{J_O} = \frac{n_O}{n_p} - \frac{C_H}{n_O n_p L_O} \times \frac{n_O L_O \Delta\mu_H}{J_O} \tag{8.13}$$

as follows from the definitions of J_p and J_O. Thus if $C_H = 0$, that is, if the protons do not return to the interior without phosphorylating, the ratio ATP/O_2 will be $n_O/n_p = 3$, as expected in the purely chemical model. Now, as C_H increases, the ratio J_p/J_O remains less than n_O/n_p and is increasingly smaller as $\Delta\mu_H$ increases. This brings out the limitations of the calculations of stoichiometry and efficiencies based on the chemical ideas of equilibrium thermodynamics.

8.5 EFFICIENCY OF THE ENERGY CONVERSION

The equations of the thermodynamics of irreversible processes allow us to also study the efficiency as a function of the proton flux. Let us define the efficiency as

$$\eta = -\frac{J_p A_p}{J_O A_O}$$

as in (6.27). This is immediately obtained explicitly by starting from (8.13), since

$$\eta = -\frac{n_0 A_p}{n_p A_0} + \frac{C_H}{n_0 n_p L_0} \times \frac{n_0 L_0 \Delta\mu_H A_p}{J_0 A_0} \tag{8.14}$$

Let us recall that $n_0 = H^+/0_2$ and $n_p = H^+/ATP$ are the number of protons expelled for each molecule of O_2 consumed and the number of protons necessary to produce one ATP, respectively. These parameters are characteristic for the pumps, but this is not so for J_p/J_0 or A_pJ_p/A_0J_0. When we recall this definition of n_0 and n_p, obviously the first term on the right-hand side of (8.14) is precisely the efficiency calculated in the classical way for equilibrium in the first section of this chapter. We see, therefore, that away from equilibrium the efficiency decreases, since A_p/A_0 is negative.

Finally, we note that the coefficients C_H, L_0, and L_p can be modified by a series of inhibitors. Thus, for example, proton carriers such as the already cited dinitrophenol or weak fatty acids increase the value of C_H, and consequently decrease the efficiency. The coefficient L_p is decreased by ATPase inhibitors like oligomycin; therefore, L_0 can be reduced by inhibitors of the oxidation chain, such as different cyanides.

In regard to the maximum efficiency and the minimum entropy production, discussed in the last chapter, we have studied how we can make the state of minimum entropy production coincide with the state of maximum efficiency. For this purpose, we assume a third reaction, regulated by the phosphorylation potential, which would consume ATP with the rhythm necessary to optimize the efficiency and minimize the entropy at the same time. This analysis allows us to give a thermodynamic interpretation to a series of enzymes regulated by ATP which appear to carry out the functions of the kind we have mentioned before. The Swiss scientist Stucki has studied the optimization of various parameters in the case of oxidative phosphorylation. Thus, the phosphorylation rate J_p is maximum when $q = 0.78$, and the power is maximum when $q = 0.91$. The values that are obtained for the phenomenological coefficients are $L_{00} = 0.616$, $L_{01} = 0.198$, $L_{10} = 0.214$ and $L_{11} = 0.076$ $(\mu\text{mole})^2 \cdot \text{cm}^{-2} \cdot \text{kcal}^{-1}$, so that $Z = 2.84$ and the degree of coupling is $q = 0.95$.

8.6 CONCLUSIONS

In this chapter, we have presented an overall formulation for oxidative phosphorylation in terms of the thermodynamics of irreversible processes. In this scheme, we have commented on the consequences of the chemical and chemiosmotic hypotheses with regard to the phenomenological coefficients. In the scheme provided by the chemiosmotic theory, we have studied the stoichiometry and the efficiency, stating the limitations of the calculations performed within the framework of the equilibrium theories: the information should be obtained by means of experimentation. The thermodynamics of irreversible processes nevertheless provides a working scheme and a theoretical framework that guides the possible questions of the experimenters, and helps to critically interpret their results. Despite the fact that by itself it cannot give an answer to the most popular questions about oxidative phosphorylation, the use of irreversible thermodynamics offers, within its limitations, a clearly satisfactory balance.

BIBLIOGRAPHY

CAPLAN, S. ROY, and A. ESSIG, *Bioenergetics and Linear Nonequilibrium Thermodynamics. The Steady State.* Cambridge, Massachusetts: Harvard University Press, 1983.

HINKLE, P. C., and R. E. MCCARTY, "How cells make ATP," *Scientific American* 283, No. 3 (1978), 104–17, 121–23.

NICHOLLS, D. G., *Bioenergetics. An Introduction to the Chemiosmotic Theory.* London: Academic Press, 1982.

STUCKI, J. W., "The optimal efficiency and the economic degrees of coupling of oxidative phosphorylation," *European Journal of Biochemistry* 109 (1980), 269–83.

NONLINEAR THEORY

9

Stability Theory

9.1 INTRODUCTION

The equilibrium state is characterized by the extremum character (maximum or minimum) of any thermodynamic potential. If a system is isolated, the equilibrium state that is reached is the one with maximum entropy. If the system evolves at constant T and p, the equilibrium state corresponds to the minimum in the free energy G. If the volume and the temperature are held fixed, the system tends toward the state of minimum free energy F, and similarly in other physical situations. The question that presents itself next is whether or not these equilibrium states are stable, that is, whether a small perturbation is capable of leading to the production of large changes in the state of the system, or if it damps out without any consequences. At the end of Chapter 7, we commented on the great importance of stability problems in the case of nerve conduction: small perturbations are damped out and large perturbations provoke transmission of the action potential.

The instability of the equilibrium states is intimately related to the phase changes, that is, to the collective reorderings and the appearance of structures in the system. Thus, for example, the equilibrium state of liquid water is no longer stable below zero degrees; the appearance of any small perturbation under these circumstances provokes a radical change in the state of the system. The liquid transforms to the solid, with the appearance of a new structure, more ordered than that of the liquid. Once this change (which we call a phase transition) has occurred, the system remains in the new structural state without the need to supply energy to it or extract energy from it, as long as it is held at constant temperature.

The situation for nonequilibrium stationary states is very similar. When a system is gradually removed from equilibrium (for example, by increasing the temperature differences at various points in the system), at some moment the corresponding stationary state is no longer stable. Once this situation has been reached, the system reorders itself: A structural "mutation" appears. In order to maintain the system in this new configuration, we need to continuously supply energy, since we are dealing with a nonequilibrium state. If we do not supply energy to the system (if the living being stops respiration), the system approaches equilibrium and the structurization disappears. For this reason, the structures that appear outside equilibrium (the ones to which we dedicate these last three chapters) have been called by Prigogine *dissipative structures*, as opposed to the conservative structures of equilibrium. As we see in the remaining chapters, these dissipative structures have an enormous importance in biology, since they explain how living bodies can be structured in time and space. Thus the old ill-posed problem is resolved, which presented structurization of living beings as in opposition and contradiction to the thermodynamic disordering of isolated systems. This problem is resolved when we note that living beings are not isolated systems and therefore their entropy can decrease, on the condition that the entropy of their environment increases sufficiently. This eliminates the contradiction with thermodynamics, but does not explain the appearance of structures. Nonlinear thermodynamics and the study of the appearance of nonequilibrium dissipative structures go much further in clarifying many aspects of this old question.

9.2 STABILITY OF EQUILIBRIUM STATES

In order to facilitate understanding of the subject of stability, let us assume a system at constant T and p. Its equilibrium state will be the one corresponding to the minimum free energy G. Let us now assume that the free energy has the form

$$G(T, p, x) = a(T, p)x^2 \tag{9.1}$$

where a is a function of T and p, and x is a parameter that characterizes the state of the system (for example, the fraction of gas). If $a > 0$, the minimum in $G(T, p, x)$ obviously corresponds to $x = 0$ and the system is characterized by not having a gaseous fraction. In this case, the equilibrium state will be stable, since $x = 0$ corresponds to a minimum in the free energy G. In other words, the second derivative of G with respect to x is positive (the condition for the minimum), since

$$\delta^2 G = 2a(\delta x)^2 > 0 \tag{9.2}$$

We have expressed as δx the possible perturbations in the system of internal origin (spontaneous fluctuations) or external origin, and $\delta^2 G$ indicates the second variation in G; that is,

$$\delta^2 G = (d^2G/dx^2)\,(\delta x)^2$$

Let us now assume that the Gibbs function has a more complicated form

$$G(T, p, x) = a(T_0 - T)x^2 + bx^4 \tag{9.3}$$

where a and b are positive constants, T is the temperature, and T_0 is a reference temperature. The extrema of this function can be given by the equation

$$dG/dx = 2a(T_0 - T)x + 4bx^3 = 0$$

that is, for $x = 0$ and for $x = \pm[(a/2b)(T - T_0)]^{1/2}$. We observe that if $T < T_0$, the argument of the square root is negative and the only extremum of the function is found at $x = 0$. On the other hand, when $T > T_0$, the free energy G has two minima, given by the two solutions of the square root, while the state characterized by $x = 0$ corresponds to a maximum (Fig. 9.1).

We can make this clear if we study the second variation in G, which is equal to

$$\delta^2 G = [2a(T_0 - T) + 12bx^2](\delta x)^2 \tag{9.4}$$

If $T < T_0$, expression (9.4) is positive and, in particular, it is positive for $x = 0$. If $T > T_0$, on the other hand, this expression is negative for $x = 0$. Thus, study of $\delta^2 G$ allows us to see that the state corresponding to $x = 0$ is no longer stable when the temperature is greater than T_0. For $T < T_0$, the entire system is found in the condensed phase (solid or liquid); while for $T > T_0$, some fraction of the system (x) is found in the gaseous state. We observe that this is not what happens with ordinary boiling, in which the system passes from the liquid phase ($x = 0$) to the gas phase ($x \neq 0$) at some given temperature T_0, in a way such that x does not increase gradually with $T - T_0$, but rather changes abruptly when the temperature T_0 is reached. This type of discontinuous transition is called a first-order transition in thermodynamics, while the continuous transition that we have described is called a second-order transition.

If we represent the value of x in equilibrium as a function of the temperature, we obtain Fig. 9.2. In this figure, we see that for $T < T_0$, the equilibrium value is $x = 0$. On the other hand, starting from $T =$

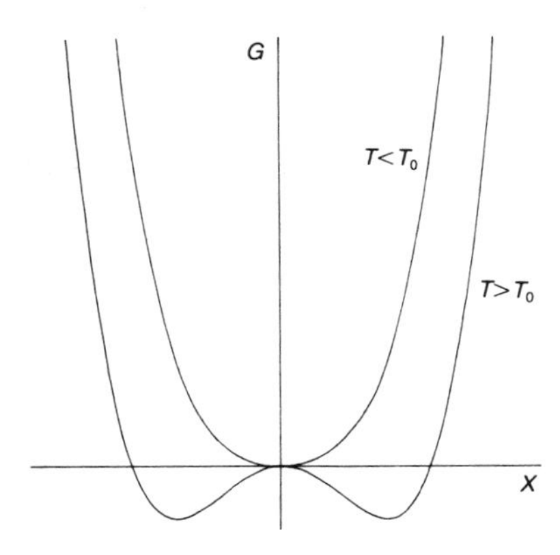

Figure 9.1 Representation of G vs. x for the function G given by Eq. (9.3).

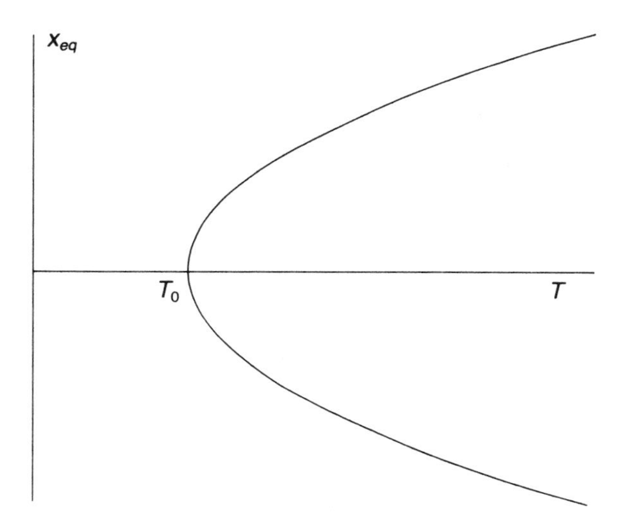

Figure 9.2 Representation of the first bifurcation.

T_0, what is called a bifurcation occurs: the equilibrium value is no longer zero, and the system experiences some disordering (or ordering), since a disordered gas phase appears.

In the case of a first-order transition, we have a reverse bifurcation of the form marked in Fig. 9.3, in which hysteresis is possible (the process

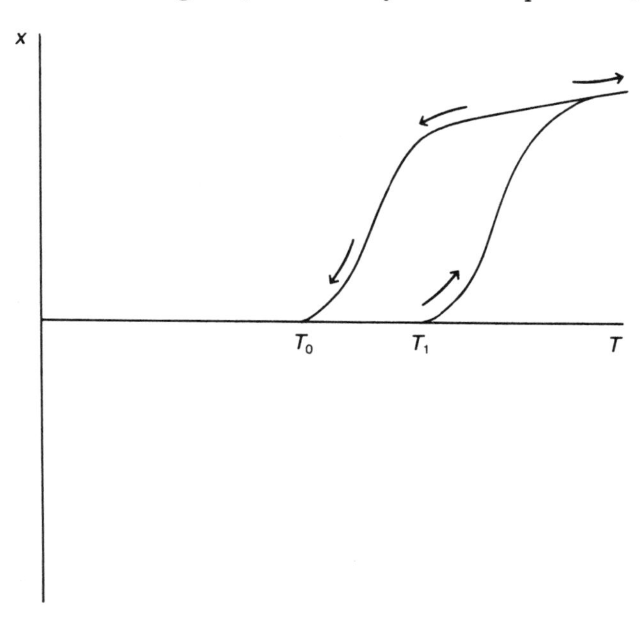

Figure 9.3 Reverse bifurcation.

following raising the temperature can be different from the process following cooling the system), a familiar phenomenon in ferromagnetism, and supercooling and metastable states are possible. These phenomena have been studied in detail within the framework of catastrophe theory.

In the heating process, when $T = T_0$ is reached, the system changes state abruptly. In cooling, the reverse change in state can occur at a different temperature T_1. For some systems, T_0 is much lower than T_1 (hysteresis), while for other systems the two temperatures virtually coincide.

To summarize, the principal ideas in this section are (1) stable equilibrium states are those that correspond to minima in the Gibbs function; (2) the necessary (but not sufficient) condition is that the first derivative of the free energy G be zero. This condition allows us to locate the possible equilibrium states. However, only those that correspond to points at which the second derivative is positive are stable. This last condition allows us to express the stability or instability of a state of the system in mathematical form.

9.3 STABILITY OF THE NONEQUILIBRIUM STATIONARY STATES

Passing from the problem of equilibrium to the problem away from equilibrium, we must establish a mathematical criterion for stability. We have already seen what is the criterion in the equilibrium case: that the second derivative of the Gibbs function be positive, or that the second derivative of the entropy be negative, depending on whether the system is found at constant T and p or the system is isolated. What is the stability criterion away from equilibrium?

To derive the answer to this question is outside the scope of this text. The mathematical theory of stability assures that a state is stable when it satisfies the following conditions:

$$\delta^2 S > 0 \tag{9.5}$$

$$d(\delta^2 S)/dt > 0 \tag{9.6}$$

In this case, we say that the function $\delta^2 S$ operates as a Lyapunov function and assures the stability of the stationary state. We observe that in equilibrium, condition (9.5) is sufficient. On the other hand, when we are dealing with a nonequilibrium situation (which, therefore, has not only energetic but also dynamic implications), it is necessary to satisfy not only condition (9.5), which refers to static factors, but also (9.6). The latter takes into account the dynamic factors, such as small modifications in the rhythm with which energy is supplied to the system—perturbations that do not make sense in an equilibrium state, where we do not need to extract or supply energy, but which can be vital in a nonequilibrium stationary state.

Let us assume that condition (9.5) is automatically satisfied, that is, that we find ourselves sufficiently far away from the phase changes in the system: in the case of water, let the temperature be between 1 and 99 degrees Celsius, for example, so that the water neither freezes nor evaporates. In turn, let us carefully examine condition (9.6). Let us recall that the time derivative of the entropy is basically the entropy production $dS/dt = \sigma$, according to (3.5). Therefore

$$d(\delta^2 S)/dt = \delta^2(dS/dt) = \delta^2\sigma > 0 \tag{9.7}$$

Starting from this criterion, let us look for the conditions under which the stationary states are no longer stable. Since the entropy production is the sum of the products of the fluxes times the forces, $\delta^2\sigma$ could be represented, in the simplest case of a single flux times a single force, that is, when $\sigma = J \cdot X$, as

$$\delta^2\sigma = \delta J \cdot \delta X \tag{9.8}$$

Here, δJ is the perturbation of the flux and δX is the perturbation of the thermodynamic force. As we have said before, these should be due both to perturbations of external origin (collisions, vibrations) and to spontaneous fluctuations of the system. If the phenomenological law is linear, that is, if $J = LX$, with $L > 0$, according to the second law we have

$$\delta^2\sigma = \delta J \cdot \delta X = L(\delta X)^2 \geqslant 0 \tag{9.9}$$

Since this quantity is always positive, conclude that the stationary states described by the linear phenomenological laws are always thermodynamically stable. This case is analogous to the equilibrium situation in which

$$G(x) = ax^2$$

studied in the second section of this chapter.

Now let us suppose that the phenomenological laws are not linear as we have assumed up to now, but that they contain nonlinear terms such as

$$J = LX - L'X^2 \tag{9.10}$$

What are the possible physical implications or consequences of these nonlinear terms? Let us examine the situation. In this case, the second variation of the entropy production will be, introducing Eq. (9.10) into Eq. (9.8):

$$\delta^2\sigma = \delta J \cdot \delta X = L(\delta X)^2 - 2L'X_0(\delta X)^2 \tag{9.11}$$

where we have taken into account the fact that $\delta(X^2) = 2X_0\delta X$, since δX, which involves small variations, behaves like an ordinary differential operator; on the other hand, x_0 indicates the value of the thermodynamic force X in the stationary state that is under study.

While in the case of linear phenomenological laws, (9.9) is always positive and the stability condition (9.6) is satisfied, in this case (9.11) can be negative when $X_0 > L/2L'$. Therefore, if the thermodynamic force (which, we recall, characterizes the degree of imbalance) increases above this critical value, that is, if the system is sufficiently far from equilibrium, the stationary state will no longer be stable. We observe that as L' (the nonlinear coefficient) becomes smaller, the value of X_0 (or the degree of imbalance necessary to destabilize the system) will become greater.

What type of phenomenon (structurization, ordering) will appear when one of these instabilities is displayed? We study this question in the next chapter, with special reference to some situations of biological interest. Summing up this chapter, we should recall that when the phenomenological laws are no longer linear, the nonlinear stationary states are no longer stable. Starting from these instabilities, the system can display definite structurizations, sometimes of great biological interest.

BIBLIOGRAPHY

GLANSDORFF, P., and I. PRIGOGINE, *Thermodynamics of Structure, Stability, and Fluctuations*. New York: Wiley, 1971.

VOLKENSHTEIN, M. V., *Biophysics*. Moscow: Mir, 1983.

10

Ordering in Time and Space

10.1 INTRODUCTION

As we have seen, the appearance of nonlinear terms in the constitutive equations of a system can lead to thermodynamic instabilities, starting from which the system can display different types of order. In this chapter, we study ordering in time (that is, the appearance of rhythms) and ordering in space (or morphological structurization).

The law of mass action implies that the equations of chemical kinetics in general are nonlinear. The equations describing interaction of various species in ecology, analogous in so many aspects to the equations of chemical kinetics, also display this characteristic. Thus, the previous study of stability of nonequilibrium stationary states is not simply academic but rather is strongly motivated by the impact of reality in chemical and biological systems.

As a consequence of a thermodynamic or hydrodynamic instability, the structurization of biological systems can be temporal and morphological. How can we arrive at the complicated structures that are commonly displayed by organs of living beings? This has been a subject which for centuries has excited the interest of scientists and the curiosity of any person with minimal talents for observation and some capacity for reflection. The mathematical theories of morphogenesis have stimulated important progress in these fields, both from the viewpoint of biology and from that of pure mathematics. It is enough to mention the example of the catastrophe theory of René Thom, which has resonated so much in so many fields of science. For the level of this book, it would not be appropriate to have recourse to sophisticated mathematical techniques. Within the scheme we have presented and in order to complete it, we limit ourselves to discussion of a few examples that allow us to intuitively understand the beginning of a path that has been shown to be very fruitful.

It is important to point out how interesting and exceptional is the appearance of spontaneous rhythms. Until relatively recently, it was believed that thermodynamics prohibited this type of behavior. Thus, for example, when we studied the equalization of temperatures of two systems in thermal contact (in Chapter 3), we saw that the process is exponential and not oscillatory. If the process of equalization of the temperatures were oscillatory, there would be moments at which heat would pass from the cold body to the hot body, contrary to the statement of the second law. We observe then that the situation in thermodynamics is different from the one presented in mechanics, where oscillating systems are so frequent. For this reason, the appearance of spontaneous rhythms in chemical and biological systems is a great surprise, and for some time no one could see how to include it in thermodynamic schemes. As we will see, it is precisely in the appearance of instabilities in nonequilibrium stationary states that this phenomenon finds the appropriate point for its insertion into thermodynamics.

10.2 ORDERING OF A CONVECTIVE FLUID: BÉNARD'S PROBLEM

The most familiar example of a dissipative structure is given by the convection of a viscous fluid. This problem (called the Rayleigh–Bénard problem in honor of the French researcher H. Bénard, who observed it for the first time at the beginning of this century, and the English physicist Lord Rayleigh, who provided the first intuitive theoretical explanation) has constituted a true scientific paradigm, in the sense that Kuhn gave this expression. In the last ten years, more than three thousand scientific articles have been published on its subtleties, and it has been the testing ground for many experimental techniques and many innovative mathematical methods. We will comment in more detail on this in the next chapter.

The simplest example of Bénard's problem consists of heating a fluid of viscosity η and density ρ held between two parallel horizontal plates separated by a generally small distance d. This heating is done through the lower plate, whose temperature is T_2 and can be regulated and varied. The temperature of the upper plate, T_1, is held constant by any cooling method. The thermal expansion coefficient of the fluid is α. When $T_1 = T_2$, the fluid is in thermal equilibrium at the temperature $T = T_1 = T_2$. In this case, its properties are uniform and do not display any kind of structure.

If we begin to heat the lower plate, that is, to increase T_2, the fluid in contact with the plate will expand and therefore will become less dense. By Archimedes's principle, it will tend to rise. This tendency toward imbalance is counteracted in principle by dissipative effects: on the one hand, the viscosity of the fluid (that is, its internal friction) exerts a force that hinders the fluid from beginning to be set in motion; on the other hand, the thermal conductivity of the fluid allows it to transfer a good

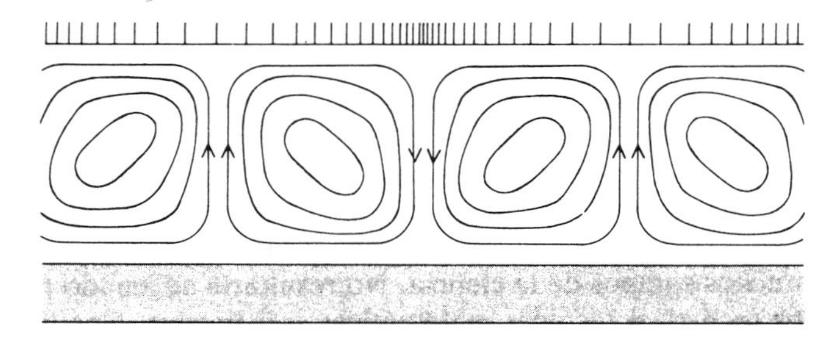

Figure 10.1 Bénard's device.

part of the heat (instead of accumulating energy and expanding), which decreases its expansion.

When the temperature difference $T_2 - T_1$ is small, the dissipative effects that oppose the motion are greater than the expansion effects, so the fluid remains immobile. Heat is transferred through it by thermal conduction, that is, without macroscopic movement of the fluid, and the fluid does not display any particular structurization except for the monotonic variation of the temperature from T_1 to T_2 from the upper layer to the lower layer.

When the difference $T_2 - T_1$ reaches a certain critical value, the elevating effects of expansion equal the dissipative effects. For values of $T_2 - T_1$ slightly greater than this critical value, the elevating effects predominate and the fluid starts to move. The curious fact is that this movement is perfectly structured: The fluid is divided into horizontal cylindrical convection cells. In these cells, the fluid rotates in a vertical plane. At some points, the hot liquid rises; once on top, it is cooled and its density increases again, inducing a movement downward.

We see how, when sufficiently far from equilibrium (that is, for sufficiently large values of the temperature difference between the plates), the system *structures itself spontaneously*. This type of spontaneous collective organization constitutes a field of study in a very recent branch of physics called *synergetics*, which includes thermodynamics, statistical mechanics, chemistry, optics, ecology, and biology. Synergetics is one of the most important multidisciplinary efforts of the last decade.

However, this structure disappears as soon as we stop heating the system since, by eliminating the supply of heat, the temperature difference decreases, the movement ceases, and the liquid finally reaches the equilibrium state of uniform temperature. In other words, we need to continuously "feed" the system in order for it to maintain its spontaneous structure. As soon as we stop "feeding" it, its structure starts to disappear, the system "dies," and it reaches the equilibrium state.

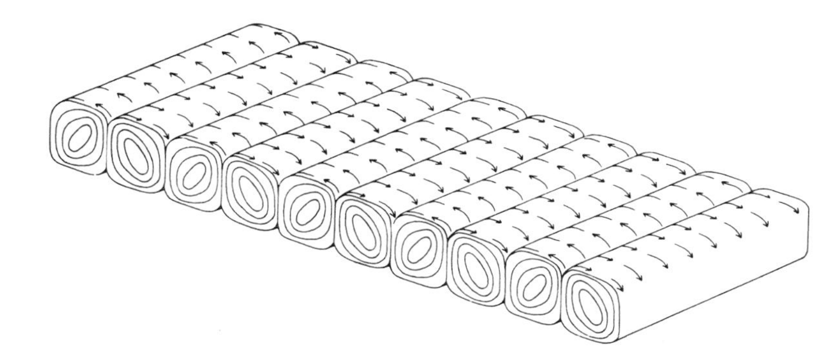

Figure 10.2 Convective cylinders in a Bénard cell.

Let us comment briefly on some quantitative aspects of this problem. What is the critical temperature difference? As we have said, it depends on a compromise between various factors. These factors are included in the dimensionless Rayleigh number Ra, defined as

$$Ra \equiv \frac{g\alpha d^3}{\eta k}(T_2 - T_1) \tag{10.1}$$

where g is the acceleration of gravity. As we can prove by an analysis of stability starting from the equations of fluid mechanics, the critical value of the Rayleigh number is $Ra_c = 1707$.

Thus the critical temperature difference is

$$(T_2 - T_1)_c = 1707\eta k/g\alpha d^3 \tag{10.2}$$

When the dissipative coefficients that oppose the motion, η or k, are large, it takes more to achieve convection than when they are small. On the other hand, when the expansion coefficient is greater, the destabilizing effects are more appreciable and the fluid achieves convection with a smaller temperature difference.

According to hydrodynamic analysis, the velocity distribution in the fluid varies periodically in space as

$$v(x) \approx [(Ra - Ra_c)/\Phi]^{1/2} \cos(2\pi x/\lambda) \tag{10.3}$$

where λ is the wavelength, that is, the repetition length of the horizontal cells, which turns out to be $\lambda \approx 2.2d$, and Φ is a constant.

If we represent the amplitude of the velocity as a function of the Rayleigh number, we have the *bifurcation* scheme of Fig. 10.3. Below the critical temperature difference, the system is not structured and the velocity is zero everywhere. When the critical temperature difference is reached, the resting state becomes unstable and the system can choose between two structured states with parallel cylinders: in one state, the

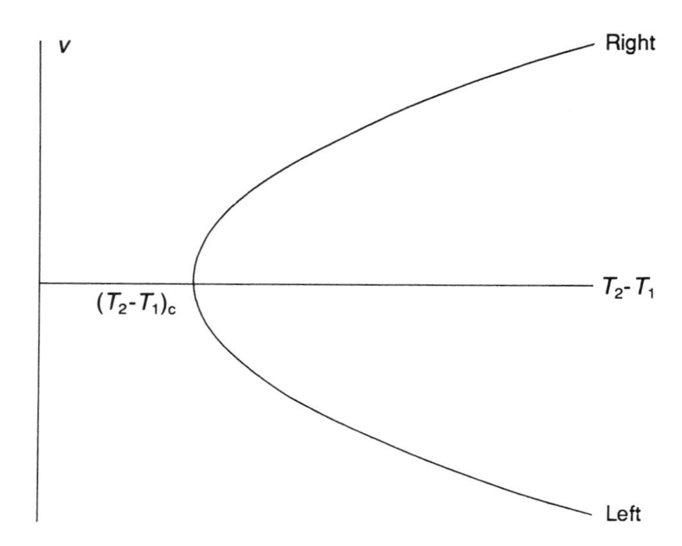

Figure 10.3 Diagram of the bifurcation in the velocity in Bénard's problem.

first cylinder rotates toward the right; in the other state, the first cylinder rotates toward the left.

The most spectacular case of convective structure is not achieved when the liquid is between two plates, but rather when its upper surface is in contact with the air. In this case, the surface tension of the liquid plays an important role and the convective cells are no longer horizontal cylinders, but rather vertical hexagonal prisms (Fig. 10.4), through whose center the cold fluid descends and through whose faces the hot liquid ascends. This phenomenon also occurs in terrestrial magma, which can

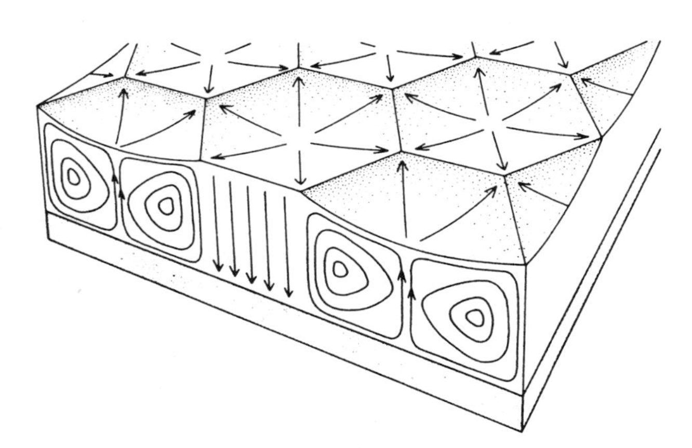

Figure 10.4 Hexagonal Bénard cells.

be modeled as a very dense and very viscous fluid, hotter below (near the center of the earth) than near the surface. Thus, the ascending movements of magma would display some polygonal structures, which in fact have been observed on a grand scale in the distribution of volcanos in the Sea of Japan.

This problem, in its simplicity, perfectly illustrates the fundamental idea of this last part: When the system is close to equilibrium, that is, when the temperature differences between the lower plate and the upper plate are sufficiently small, the stationary state of the system is characterized by a simple variation in temperature as a function of position. Then the system does not display any special structurization. However, when the system is sufficiently far from equilibrium, that is, when the temperature difference reaches a critical value, structured convective movement appears. In other words, removal from equilibrium is an ordering or structurizing factor that facilitates a "mutation" from a low-order state, relatively homogeneous, to a more ordered state. In the language of physics, we would say that we have a "breaking of symmetry." The homogeneity of the horizontal translation has been lost; the structure is no longer homogeneous, but rather periodic.

Despite the fact that this example is not related directly to biology, its clarity is inescapable and it brings out the existence of two zones in the study of the thermodynamics of irreversible processes: one refers to states close to equilibrium, without any special order, while the other deals with typical ordering phenomena of states sufficiently far from equilibrium.

10.3 SPATIAL ORDERING IN CHEMICAL REACTIONS

The convective case is not the most appropriate one with regard to biological structurization problems. The greater part of biological mechanisms are of the chemical type. In this section, we see how, in these cases also, the phenomenon of spontaneous ordering can occur when the system is sufficiently far removed from equilibrium.

The most familiar theoretical model in this field is called the Brusselator, a scheme of reactions proposed by Prigogine and Lefever in 1968, which allows us to explain phenomena observed in various chemical reactions. The scheme consists of the following reactions:

$$\begin{aligned} A &\xrightarrow{k_1} X \\ B + X &\xrightarrow{k_2} Y + D \\ 2X + Y &\xrightarrow{k_3} 3X \\ X &\xrightarrow{k_4} E \end{aligned} \tag{10.4}$$

Overall, the equation is reduced to $A + B \longrightarrow E + D$, while X and Y are intermediates on which we will focus our attention. Let us assume a system at rest in which the reactants A and B are held at constant concentration; either they are consumed very slowly or they are supplied continuously. The products D and E are withdrawn from the system. The compounds X and Y, on the other hand, cannot cross through the walls of the system, but they can diffuse into its interior. The equations which, according to the law of mass action, describe the change in the concentrations of X and Y are

$$(\partial X/\partial t) = A - (B + 1)X + X^2Y + D_X(\partial^2 X/\partial r^2) \qquad (10.5)$$

$$(\partial Y/\partial t) = BX - X^2Y + D_Y(\partial^2 Y/\partial r^2) \qquad (10.6)$$

We have assumed, for simplification, that the rate constants are unity. D_X and D_Y, on the other hand, are the respective diffusion coefficients, and r is the distance to the origin (we assume a problem in which only one dimension is relevant, which is the case in a test tube or in a worm).

This system has a uniform steady-state solution given by $X_0 = A$, $Y_0 = (B/A)$. This solution, however, is stable only for the concentration of B which is less than some critical value B_c given by

$$B_c = A^2 + 1 + (m\pi/l)^2(D_X + D_Y) \qquad (10.7)$$

where l is the length of the system and m is some integer. If we gradually increase the concentration of B, for $B < B_c$ there is a uniform state; but when $B = B_c$, the system passes to a nonuniform state in which the concentration of X, for example, varies with position according to

$$X = A \pm [(B - B_c)/\Phi]^{1/2} \cos(m\pi r/l) + O(B - B_c) \qquad (10.8)$$

The first value of B for which this can be observed is the one corresponding to $m = 1$. The concentration of X varies over space, as shown in Fig. 10.5. In order to understand how exceptional this phenomenon is, consider that if we had only diffusion, the system would pass rapidly from an inhomogeneous state to a homogeneous state. Here, on the other hand, because of the nonlinear terms in the equations (that is, because of the third reaction in (10.4), the autocatalytic reaction), the system, which is originally well mixed in a uniform state, orders itself spontaneously into one zone rich in X and one zone poor in X.

The importance of this phenomenon especially lies in the possibility of explaining some simple processes of morphogenesis or cell differentiation. In the beginning, the organism is unicellular, and during gestation or growth this single cell is replaced by many differentiated cells. How does this differentiation occur? In the case of the hydra, to give a concrete and very simplified example, we could distinguish between the body and the head. (In reality, there are more different types of cells. But here we

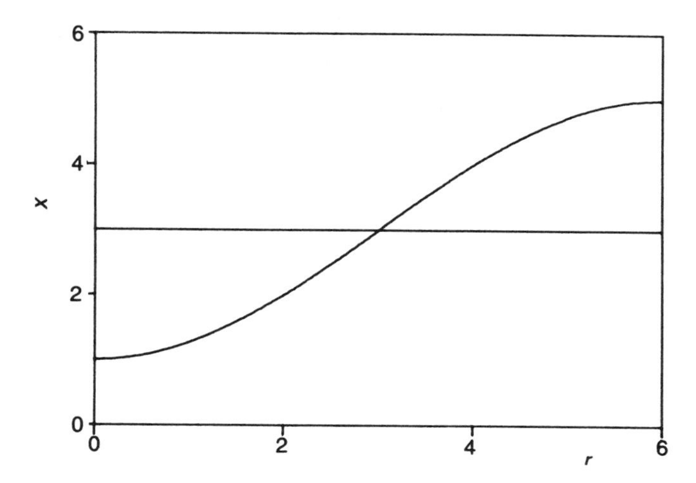

Figure 10.5 Variation in X with respect to $r \cdot m = 1$.

will concentrate on only two.) A certain substance has been discovered (a low-molecular weight polypeptide) which, starting from a given concentration, induces this cellular differentiation by regulating the reading of the appropriate zone by the DNA; the cells of the body are transformed, so to speak, into cells of the head. The problem is: Why does this substance accumulate spontaneously at only one end of the individual? We have seen one possible answer in the Brusselator model. We also note that if l, the length of the system, is very small, the critical concentration of B given by Eq. (10.7) will be very high; that is, it will be very difficult for the system to achieve it. So the uniform state will always be stable and will not produce differentiation. Thus, hydras begin to have a differentiated head only when they exceed a certain critical length, a fact that is well-known to zoologists.

We observe that, according to Eq. (10.8), we could also have spontaneous structures with $m > 1$. If m were equal to 5, for example, we would have a spatial distribution of X of the form represented in Fig. 10.6. The system would undergo self-organization, passing from a uniform distribution of concentrations to a banded distribution.

There are reactions that display very spectacular behavior, like the Belousov–Zhabotinskii reaction, which we will discuss in the next section. If we add colored substances whose color varies according to the pH or according to some other factor (the concentration of X, for example), this reaction also organizes itself into bands (Fig. 10.7).

The phenomenon of spontaneous structurization, with accumulation of a certain substance at a certain point, is the basis for the organized behavior of some insects (ants, termites). When such insects excavate or

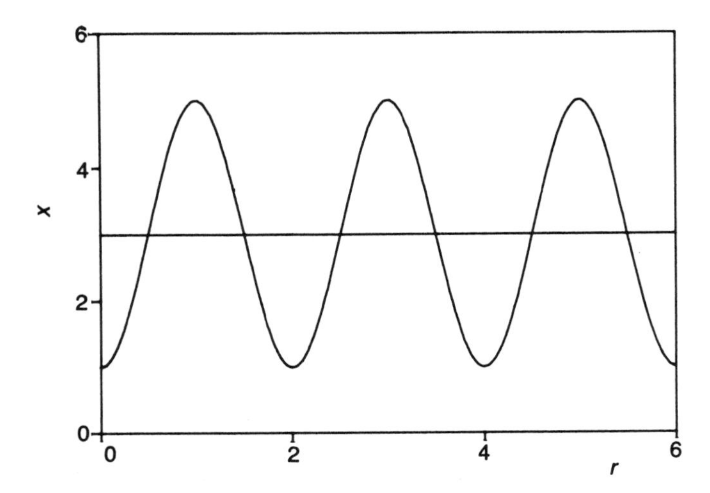

Figure 10.6 Variation in X with respect to $r \cdot m = 6$.

construct their nest, each individual leaves soil transported to some point on a territory and mixed with pheromones, an attractive substance. Starting from a critical pheromone concentration, a phenomenon of spontaneous organization occurs and all the individuals leave soil or excavate

Figure 10.7 Belousov–Zhabotinskii reaction.

at the same point. The high concentration of pheromones at that point will attract increasingly more individuals, so the corresponding mound of soil will grow rapidly.

10.4 ORDERING IN TIME. BIOLOGICAL CLOCKS

In addition to producing organization in space, moving the system away from equilibrium can produce organization in time, or spontaneous rhythmic behavior. As we have already commented earlier, it seems strange that this can happen when we think in terms of the second law.

Hence the great surprise when, at the beginning of the century, chemical reactions could be observed which spontaneously display oscillatory behavior. We first study a relatively simple example that is well-known in chemistry and ecology: the Lotka–Volterra model of the predator–prey interaction.

The Lotka (1920)–Volterra (1923) model describes the interaction between two chemical or ecological species in the form

$$J_1 = dX_1/dt = k_1X_1 - k_2X_1X_2 \tag{10.9}$$

$$J_2 = dX_2/dt = k_3X_1X_2 - k_4X_2 \tag{10.10}$$

where X_1 and X_2 represent the number of individuals of species 1 and 2 respectively. Species 1 could be, for example, a fish that feeds directly on plankton, and k_1 would be its biotic potential (birth rate minus death rate). Species 2 could be a type of fish that feeds exclusively on fish of species 1, with k_4 its biotic potential. The k_2 and k_3 terms represent the interaction between both populations. When there are more individuals X_1 and more individuals X_2, then encounters between individuals of both species become more probable. Many such encounters end with the disappearance of species 1 into the digestive system of species 2. Therefore, in Eq. (10.9), the interaction term has a negative sign, since it describes the decrease in the population X_1 due to these sad events. On the other hand, the corresponding term is favorable to species 2, since the more its members can eat, the more they can reproduce and increase. Therefore, this term is positive in (10.10). We observe that if at some moment the individuals of species 1 disappear, that is, $X_1 = 0$, then species 2, without food, will disappear at the end of some time period determined by the reciprocal of the constant k_4.

These equations, in which $dX_1/dt = J_1$ and $dX_2/dt = J_2$ represent the thermodynamic fluxes, have two stationary solutions. The first such solution, trivial and uninteresting, is $X_1 = X_2 = 0$; that is, there is no species 1 or species 2. The second solution is

$$\begin{aligned} X_{10} &= (k_4/k_3) \\ X_{20} &= (k_1/k_2) \end{aligned} \tag{10.11}$$

The problem we pose next is whether this stationary state, in which $J_1 = J_2 = 0$, is stable. Let us discuss this in some detail.

If we add small perturbations δX_1, δX_2 to X_{10} and X_{20} so that $X_1 = X_{10} + \delta X_1$ and $X_2 = X_{20} + \delta X_2$, these obey the following linearized equations obtained by introducing the expressions for X_1 and for X_2 into (10.9) and (10.10) and by neglecting the terms containing products of δX_1 times δX_2,

$$d(\delta X_1)/dt = k_1 \delta X_1 - k_2(X_{10}\delta X_2 + X_{20}\delta X_1) \qquad (10.12)$$

$$d(\delta X_2)/dt = k_3(X_{10}\delta X_2 + X_{20}\delta X_1) - k_4 \delta X_2 \qquad (10.13)$$

Taking (10.11) into account, we can write these equations in the form

$$d(\delta X_1)/dt = -k_2 X_{10} \delta X_2 \qquad (10.14)$$

$$d(\delta X_2)/dt = k_3 X_{20} \delta X_1 \qquad (10.15)$$

If we now differentiate the first equation with respect to time and replace $d(\delta X_2)/dt$ by the value given in the second equation, we have

$$d^2(\delta X_1)/dt^2 = -k_2 k_3 X_{10} X_{20} \delta X_1 = -k_1 k_4 \delta X_1 \qquad (10.16)$$

The solution to this equation, which has the form corresponding to the motion of a harmonic oscillator, is a harmonic oscillation

$$\delta X_1 = \delta X_1(0) \cos(2\pi f t) \qquad (10.17)$$

where f is the frequency given by

$$f = (1/2\pi)(k_2 k_3 X_{10} X_{20})^{1/2} = (1/2\pi)(k_1 k_4)^{1/2} \qquad (10.18)$$

The value of f can be obtained directly if we use the analogy of (10.16) to the equation of harmonic motion of a spring.

Thus the stationary state (10.11) is not stable; a small initial perturbation $\delta X_1(0)$, $\delta X_2(0)$ does not tend to be cancelled out in a way such that the system returns to the initial state, but rather it persists in oscillating form in a period that can be calculated starting from Eqs. (10.17) and (10.18) and which, as we have seen, is $T = 2\pi/(k_1 k_4)^{1/2}$.

This oscillating behavior (Fig. 10.8) has a rather clear intuitive explanation: As X_1 increases, species 2 has more food and tends to increase its population. Now when X_2 increases, the number of individuals of species 1 that are devoured by those of species 2 also increases. Consequently, X_1 begins to decrease. As X_1 decreases, the amount of food available for species 2 decreases, so that X_2 decreases. As the number of predators X_2 decreases, the prey X_1 recover and increase their population, coming back to start the cycle again.

Volterra became interested in these equations based on the observations of Adriatic fishermen, who had noticed a periodicity in the captures of different species. As we have seen, the mathematical model pre-

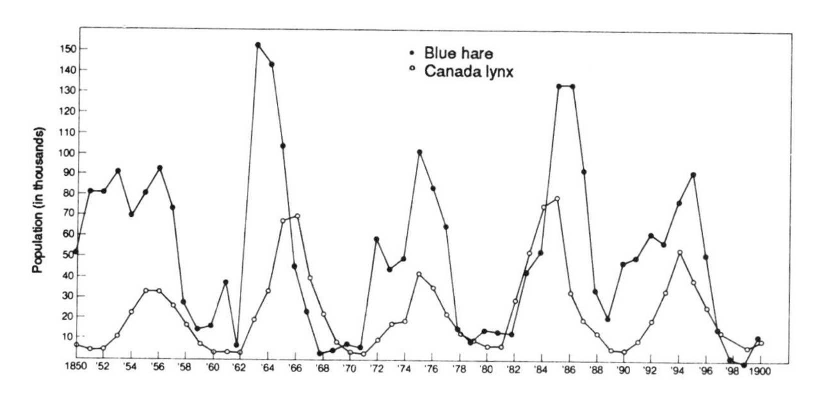

Figure 10.8 Time evolution of two predator–prey populations.

sented here allows us to justify this intuitive process and to calculate the rhythm of the ecological process in terms of the interaction between the fish and in terms of the respective biotic potentials.

The example of the oscillating reaction that we have just outlined, based on the Lotka–Volterra model, is only one particular case and moreover is not sufficiently illustrative. In fact, the oscillation characteristics (amplitude, period) depend on the initial state in which the perturbation has found the system. This is evident in the linearized model that we have presented, and we can verify that it is also true in the more general case. For this purpose, if X_{10} and X_{20} are the stationary values of X_1 and X_2, α and β are defined as $\alpha \equiv \ln(X_1/X_{10})$ and $\beta \equiv \ln(X_2/X_{20})$, and the Lotka–Volterra equations can be written as

$$(1/k_2)(d\alpha/dt) = X_{20}(1 - \exp[\beta])$$

$$(1/k_3)(d\beta/dt) = -X_{10}(1 - \exp[\alpha])$$

Multiplying the first of these equations by $X_{10}(1 - \exp[\alpha])$ and the second by $X_{20}(1 - \exp[\beta])$ and adding them, we have

$$(1/k_2)(d\alpha/dt)X_{10}(1 - \exp[\alpha]) + (1/k_3)(d\beta/dt)X_{20}(1 - \exp[\beta]) = 0$$

or equivalently

$$\frac{d}{dt}\left[\frac{X_{10}}{k_2}(\exp[\alpha] - \alpha) + \frac{X_{20}}{k_3}(\exp[\beta] - \beta)\right] = 0 \qquad (10.19)$$

Therefore, the expression between the brackets in this last equation is a constant of the motion, which depends only on k_2, k_3, and on the initial conditions $X_1(0)$ and $X_2(0)$ of the perturbation. Therefore, once these conditions are fixed, the system will move in a specific and invariant orbit.

In dissipative systems, we more frequently encounter another type of behavior called the *limit cycle*. In this case, regardless of the initial conditions, the system tends toward a certain orbit, that is, toward a certain oscillation. Figure 10.9 shows the difference between the two behaviors. In Fig. 10.9(a) we see that for each initial state, the system describes a given orbit depending on this state. On the other hand, the orbit that describes the system in the limit cycle situation [Fig. 10.9(b)] is independent of the initial state (1 or 2) of the system after a short transient period.

An example of limit cycle behavior is presented by the set of equations called the Brusselator. The model, whose spatial properties we have discussed in the last section, can represent a limit cycle. Rewriting Eqs. (10.5) and (10.6), only taking into account the variation of the concentrations of X and Y, we obtain

$$dX/dt = k_1A - k_2BX + k_3X^2Y - k_4X \tag{10.20}$$

$$dY/dt = k_2BX - k_3X^2Y \tag{10.21}$$

The stationary solution can be given by

$$X_0 = A(k_1/k_4), \qquad Y_0 = (B/A)(k_2k_4/k_1k_3)$$

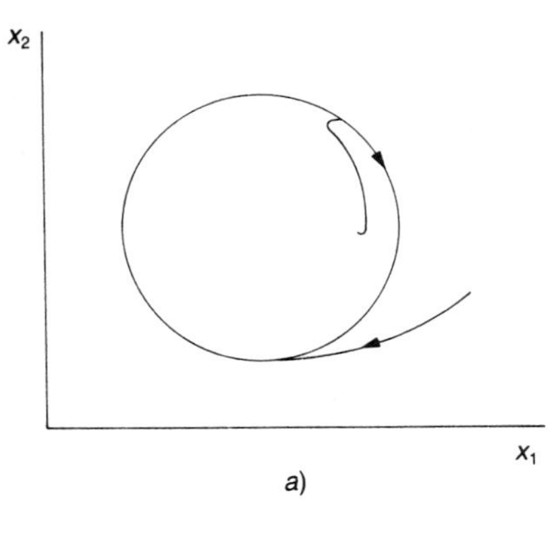

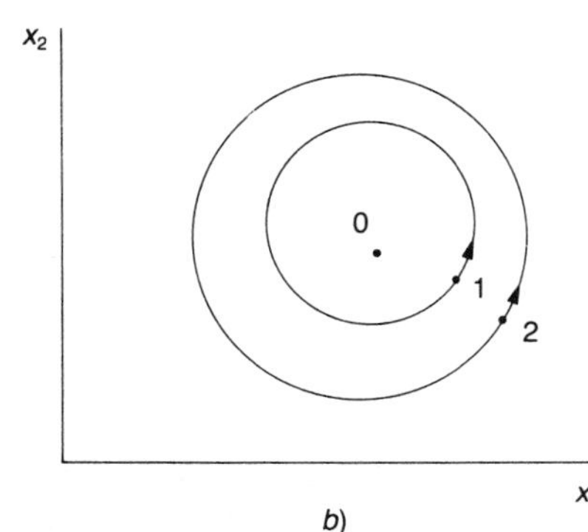

Figure 10.9 Representation of the trajectory of a system in (a) an equation of the Lotka–Volterra type; (b) a limit cycle. (0 = stationary state; 1 and 2 = different initial states)

If we now assume perturbations of the form

$$X = X_0 + \alpha \exp[\lambda t]$$
$$Y = Y_0 + \beta \exp[\lambda t] \tag{10.22}$$

and if we introduce these into Eqs. (10.20) and (10.21), we obtain the following if we neglect the terms quadratic in α and β:

$$\alpha\lambda = [B(k_2/k_4) - 1]\alpha + A^2(k_1/k_4)^2(k_3/k_4)\beta \tag{10.23}$$

$$\beta\lambda = -B(k_2/k_4)\alpha - A^2(k_1/k_4)^2(k_3/k_4)\beta \tag{10.24}$$

For α and β, the initial perturbations, to be different from zero, the determinant of the last system of equations should be equal to zero, which leads to the equation

$$\lambda^2 + (a^2 + 1 - b)\lambda + a^2 = 0 \tag{10.25}$$

where $a \equiv A(k_1/k_4)(k_3/k_4)^{1/2}$ and $b \equiv Bk_2/k_4$.

Now various situations can present themselves. If the solutions to (10.25), λ_1 and λ_2, have a part that is real and positive, then the perturbations grow and the system is unstable. If the real parts of λ_1 and λ_2 are negative, then the perturbation decreases and the system is stable. The solutions to (10.25) are

$$\lambda_{1,2} = -(\tfrac{1}{2})(a^2 + 1 - b) \pm [(a^2 + 1 - b)^2 - 4a^2]^{1/2}$$

The result is that the system will be stable if $b < a^2 + 1$ and it will be unstable if $b > a^2 + 1$. We observe that when $b = a^2 + 1$, λ_1 and λ_2 are purely imaginary, which leads to an undamped oscillation of the perturbations. Likewise, we note that for values of b between $a^2 + 2a + 1 > b > a^2 - 2a + 1$, λ_1 and λ_2 have a nonzero imaginary part, which leads to oscillatory behavior. These oscillations damp out in the stable zone ($a^2 + 1 > b > a^2 - 2a + 1$) and grow in the unstable zone ($a^2 + 2a + 1 > b > a^2 + 1$).

For values of b less than $a^2 - 2a + 1$, the perturbations decrease purely exponentially (λ_1 and λ_2 are negative and purely real), while for values of b greater than $a^2 + 2a + 1$ the perturbations increase exponentially, in principle. In fact, in practice this growth is not exponential, since as the perturbations increase it is no longer valid to neglect the appropriate terms in (10.20) and (10.21) in which these perturbations appear. It is when we take these nonlinear terms into account that the limit cycle behavior appears.

In summary, when $b = a^2 + 1$, the oscillations of the system remain fixed by the initial state, as in the Lotka–Volterra model; on the other hand, for $b > a^2 + 2a + 1$, limit cycle oscillations appear which are independent of the initial values of the perturbations in X and Y.

Finally, we note that b and a can be easily regulated just by varying the concentrations of reactants B and A respectively. The system can be maintained in situations with A, B, D, and E constant if we continually add reactants A and B and eliminate the products D and E at the appropriate rate.

The best-known oscillatory reaction is the Belousov (1958)–Zhabotinskii (1964) reaction. This reaction is a set of oxidation–reduction reactions. In one reaction, the oxidizing agent is bromate (BrO_3^-); the reducing agent is malonic acid [$H_2C(COOH)_2$], and cesium ions are used as the catalyst. The concentration of Ce^{3+} and Ce^{4+} will vary periodically with a frequency on the order of 0.01 Hz.

The overall reaction is very complicated, and its oscillatory effects can be understood if we focus only on two steps:

$$Ce^{3+} \xrightarrow{BrO_3^-,\ H^+} Ce^{4+} \qquad \text{(I, oxidation)}$$

$$Ce^{4+} \xrightarrow{\text{Malonic acid}} Ce^{3+} \qquad \text{(II, reductions)}$$

At the beginning, the system has a certain amount of Ce^{4+} ions. The second reaction provides Br^- ions, which inhibit the first reaction. Thus the concentration of Ce^{3+} increases and the solution takes on a certain color. At some time the concentration of Ce^{3+} is so great that reaction I begins. Since little Ce^{4+} remains, the system can no longer produce sufficient Br^- to inhibit the reaction and the Ce^{3+} decreases rapidly, providing Ce^{4+}, until the cycle returns to the beginning.

In biological systems we observe, from the macromolecular to the population level, oscillations in characteristic physical parameters such as enzyme activity, metabolite concentrations, parameters which govern physiological behavior, population size, and so forth.

There are numerous arguments in favor of the viewpoint that a biological system not only can but should be oscillatory. For example, let us cite the arguments considering the fact that a complex open system (which often implies autocatalytic chemical reactions like the Lotka–Volterra reaction) when far away from equilibrium evolves toward a limit cycle and consequently self-oscillations are rather probable.

Another argument considers that biological evolution has taken place on a planet, the Earth, which moves around the Sun and which rotates about its own axis. These periodicities, which induce rhythms in the changes in temperature, light, humidity, and so, have been reflected in the physiology of both animals and plants. The periodicity of day and night is clearly characteristic of living processes. In reference to this, the concept of biological clocks has been introduced, along with the name *circadian rhythms* for oscillations with a period of about 24 hours (for example, the sequence of periods of sleep and activity). The circadian rhythms are endogenic (generated internally); the physiological factor responsible for such rhythms is certainly based on self-oscillatory processes, since their periods are practically independent of temperature and other environmental factors. As other examples, we can cite some deep-sea fish which ascend during the day and descend during the night, following a 24-hour rhythm. *Penicillium diversum* exhibits a growth and sporulation rhythm with an approximate period of 24 hours, and no influence of the luminosity has been verified. *Nectria cinnabarina*, as ascomycete, exhibits a periodicity of 16 hours in growth and sporulation (when it grows on a disk, its colonies display shapes of concentric circles), with greater or lesser density of cells according to the moment at which growth has occurred. In general, the period is on the order of 24 hours, but not exactly. The etymology of its Latin name comes from this lack of exactness: *circa-die*, approximately one day. Circadian behaviors, biological clocks, are involved in the duration of the different phases of the life of insects. There are flowers, *Kalanchoe blossfeldiana*, which open and close every 23 hours. If we illuminate them with red light for a few hours, we can alter their normal rhythm; however, it takes four days to recover. Biological clocks are involved in mitosis or cell division. On the other hand, their role is not very clear in other periodic phenomena like the menstrual cycle, which involves very diverse processes operating cooperatively.

The advantages that can be provided by adaptation of the internal clock to external rhythms (diurnal, climatic) have not yet been sufficiently studied, but certainly they are of great interest.

In Chapter 6 we looked at nonlinear biochemical processes: cooperative enzyme reactions. In that case, we studied under what conditions the chemical processes, obeying the law of mass action and in general nonlinear, could be linearized. However, in general, enzyme processes are nonlinear. A typical process of this kind is

$$\begin{aligned} X + Y' &\xrightarrow{E} X' + Y \\ Y' &\longleftrightarrow Y \end{aligned} \tag{10.26}$$

The reaction is catalyzed by the enzyme E and is inhibited by the active form of the coenzyme Y'. Under certain conditions, the system is characterized by unstable stationary states and can be oscillatory.

Periodic self-oscillatory processes are characteristic of the processes of glycolysis (anaerobic catabolism of glucose) and in the conversion of ADP to ATP.

Analyzing the simplified glycolysis scheme in Fig. (10.10), if we denote X the concentration of glucose, for example, and use Y for the concentration of any intermediate (all the intermediate concentrations are proportional to one another), we obtain a set of kinetic equations

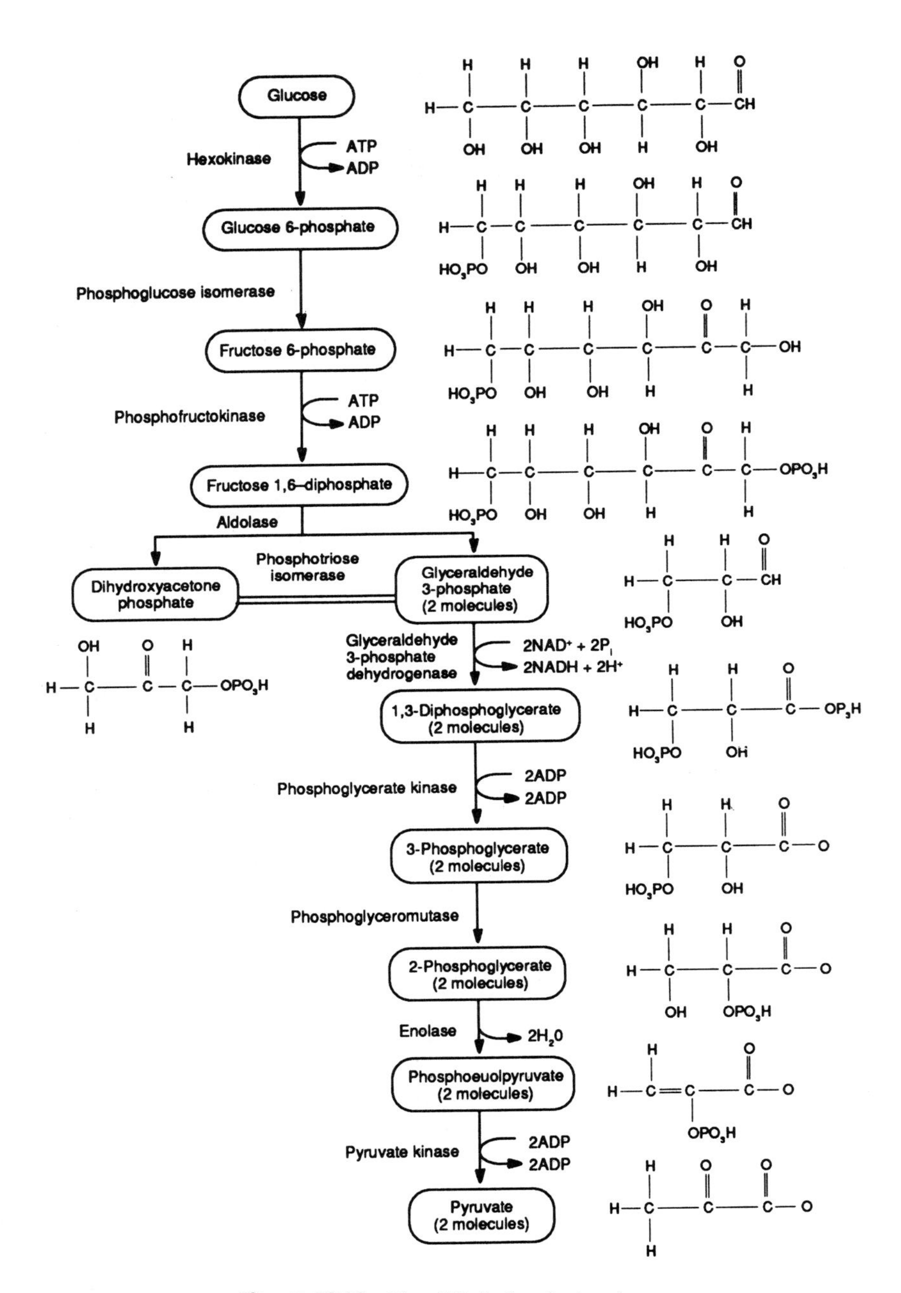

Figure 10.10 Simplified glycolysis scheme.

which, applying a series of restrictions, can be written

$$\begin{aligned} \dot{X} &= v_0 - k_1'XY \\ \dot{Y} &= k_1'XY - k_2'Y/[1 + (Y/X)] \end{aligned} \tag{10.27}$$

where v_0 is proportional to the hexokinase concentration, k_1' is proportional to the phosphofructokinase concentration, and k_2' is proportional to the pyruvate kinase content.

The steady-state concentrations of X and Y, which correspond to $\dot{X} = \dot{Y} = 0$, are

$$X(0) = \frac{k_2' - (v_0/k)}{k_1'}$$

$$Y(0) = \frac{v_0}{k_2' - (v_0/k)}$$

and defining

$$x \equiv (X/X_0), \qquad y = Y/Y_0, \qquad \gamma = k_1'v_0t/[k_2' - (v_0/k)]$$

we find that the equations in (10.27) take on the form

$$\begin{aligned} dx/d\gamma &= 1 - xy \\ dy/d\gamma &= qy[x - (1 + r)/(1 + ry)] \end{aligned} \tag{10.28}$$

where

$$q = \frac{(k_2' - v_0/k)^2}{v_0k_1'}, \qquad r = \frac{Y_0}{k}$$

Analyzing the stability of the system described by Eqs. (10.28), we see that this system spontaneously becomes oscillatory as q increases. Increment in the parameter q contributes to the amplification of the self-oscillations, and decrease in q contributes to its damping. From the expression for q it follows that the decrease in k_1' and v_0 should cause spontaneous excitation of the oscillations. We also obtain the same result when k_2' increases, that is, by adding pyruvate kinase, in accord with experiments.

In some membranes, *the excitable membranes*, the transport properties are nonlinear and can be described by equations similar to those for the enzyme reactions. For example, the Michaelis–Menten law is often used to describe facilitated transport. Just as some reactions display oscillatory behavior, under some circumstances oscillation is also possible in the process of transport across membranes. Phenomena of this type have been observed in cardiac muscle where, depending on their period, they can be very dangerous, since they can produce arrhythmias. That is, with their external rhythms, they can perturb the synchronization of wave propagation in the muscle. Multiplication of processes of this type plays an important role in some quantitative descriptions of fibrillation.

10.5 FINAL COMMENTS

The ordering of systems far from equilibrium is a very diverse subject. Modern thermodynamics has not yet sufficiently adapted to the study of this subject, which has occurred in fragmented form based on hydrodynamics, chemical kinetics, electrodynamics, and other specific dynamic theories.

While these studies were initiated in large part by Prigogine's group in Brussels, there are other very different schools and trends. The latter have grown energetically during recent years. We might cite Haken's synergetics or Fröhlich's theory of coherent excitations, to give two of the more relevant examples (both of interest in biology).

BIBLIOGRAPHY

EPSTEIN, I. R., K. KUSTIN, P. DE KEPPER, and M. ORBAN, "Oscillating chemical reactions," *Scientific American* 248, no. 3 (1983), 112–118, 120, 122–123, 146.

FRÖHLICH, H., and F. KREMER, eds., *Coherent Excitations in Biological Systems.* Berlin: Springer, 1983.

GWINNER, E., "Internal Rhythms in Bird Migration," *Scientific American* 254 (April 1986), 84–92.

HAKEN, H., *Synergetics: An Introduction to Nonequilibrium Phase Transitions and Self-Organization in Physics, Chemistry and Biology*, 2nd ed. Berlin: Springer, 1978.

KATIME AMASHTA, I. A., J. A. PEREZ-ORTIZ, A. GUTIERREZ TERRON, and F. M. GOÑI, *Termodinámica de los Procesos Irreversibles. Reacciones Oscilantes* [*Thermodynamics of Irreversible Processes. Oscillating Reactions*]. Bilbao: Universidad del Pais Vasco, 1984.

NICOLIS, G., and I. PRIGOGINE, *Self-Organization in Nonequilibrium Systems: From Dissipative Structures to Order Through Fluctuations.* New York: Wiley, 1977.

THOM, R., *Stabilité Structurelle et Morphogenèse* [*Structural Stability and Morphogenesis*]. Paris: Dunod, 1973.

VELARDE, M. G., and C. NORMAND, "Convection," *Scientific American* 243 (July 1980), 54–68.

VOLKENSHTEIN, M. B., *Biophysics* [English translation from Russian]. Moscow: Mir, 1983.

11

Farther from Equilibrium. Routes to Chaos

11.1 INTRODUCTION

In Chapter 10, we saw that, sufficiently far from equilibrium, the system arrives at a bifurcation of states, at which it passes through some temporally and spatially well-ordered structures. If the system continues to move away from equilibrium, the complexity of these structures continually increases until a point is reached where it is difficult to recognize any structure, temporal or spatial. We find ourselves in a situation called *chaotic*. In this case, we are not referring to microscopic chaos (on the molecular scale), but to macroscopic chaos.

One of the most exciting new developments for physicists in the last ten years has been the discovery of certain regularities in what until recently has been interpreted as chaos. In this chapter, we present some of these, in reference to both temporal behavior and geometry.

11.2 BIFURCATION CASCADES. A ROAD TO CHAOS

We have seen that when we increase some parameter β which describes the distance from equilibrium (a temperature or concentration gradient, the concentration of a certain reagent, for example), a value β_1 is reached which determines a situation characterized by a well-defined frequency or wavelength. Continuing to increase the parameter β, we reach a value β_2 at which the state characterized by a single frequency becomes unstable; in its place, a state appears that is characterized by two incommensurable frequencies f_1 (the initial frequency) and f_2 (the new frequency). This situation can be repeated many times for the values β_3, β_4, $\beta_5, \ldots, \beta_n, \ldots$, in such a way that we go through increasingly complicated states.

An example of this behavior is encountered in the Brusselator model of chemical reactions, discussed in Chapter 10, in which as the concentration of one of the reactants increases more and more wavelengths can appear. Also in the case of Bénard's problem, as the temperature difference between the plates that bound the fluid is increased, first some cylinders of a given diameter appear. When a second critical temperature is reached (the parameter β, in this case), these cylinders (viewed from above) adopt a sinusoidal form, with a characteristic wavelength; and so on. The scheme of this type of phenomena, or a bifurcation cascade, is what appears in Fig. 11.1.

For a long time it was believed that the critical values β_1, β_2, β_3, β_4, and so on of the parameter which represents the distance to equilibrium (reagent concentration, temperature gradient, energy flux, . . .) at which the first, second, third, . . . bifurcations occur were specific properties of each system, without any globalizing characteristic.

However, in 1975, Mitchell Feigenbaum described in an historic article (by the way, two prestigious journals rejected it for publication!) the abstract relation of very general character:

$$\lim_{n \text{ large}} \frac{\beta_n - \beta_{n-1}}{\beta_{n+1} - \beta_n} \longrightarrow 4.669201 \ldots \tag{11.1}$$

and

$$\lim_{n \text{ large}} \frac{\epsilon_n}{\epsilon_{n+1}} \longrightarrow 2.5029 \ldots \tag{11.2}$$

where ϵ_1 is the separation between the two branches coming from the $(i - 1)$th bifurcation. Thus the scheme of the bifurcation cascade displays a series of universal characteristics, that is, characteristics that are independent of the specific system.

In the following, we look at a concrete example of such a bifurcation

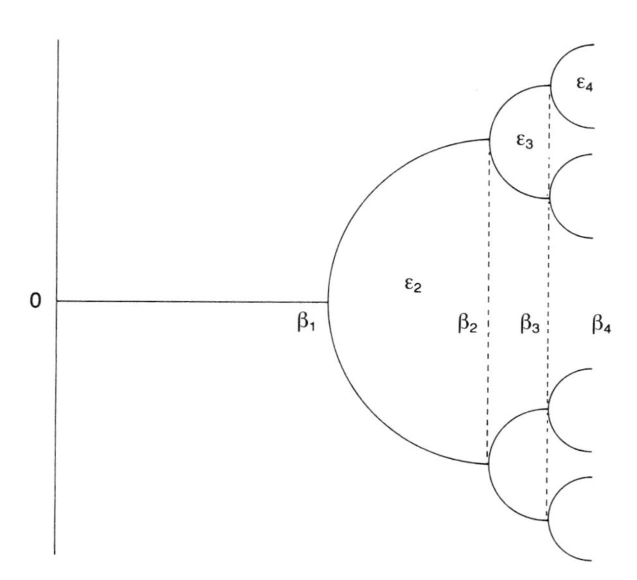

Figure 11.1 Bifurcation cascade.

cascade in a discrete dynamic system, which will show how there can be a great deal of complexity behind very simple mathematical operations. Let us suppose that there exists a single variable x whose value in time can be determined by a simple function, as

$$x_{n+1} = \beta x_n[1 - x_n] \tag{11.3}$$

where β is a parameter greater than 1 and less than 4. The values of x_n could correspond, for example, to the number of individuals of a biological species in a territory in the year n. For a small value of β, by simple calculations we can see that x_n for large n tends toward a constant value x^* given by

$$x^* = (\beta - 1)/\beta$$

independently of the initial value of x_1. When n increases, all the trajectories approach this limiting value. Since all the possible sequences of values of x_n appear to converge to this value, it is called an *attractor*. In the biological example, x^* would be the natural level to which the population tends, regardless of the initial level. If at some time many individuals die, x_n decreases abruptly but will tend to recover and return to the natural level x^*. If, on the contrary, x is greater than x^*, the population is greater than the natural level; then (perhaps because of scarcity of food or space or other environmental factors) the value of x_n decreases with time until the value x^* is reached.

The solution $x^* = (\beta - 1)/\beta$ is no longer stable when $\beta \geqslant 3$. It is easy to see the reason for the instability. Write the recurrence relation (11.3) as

$$x_{n+1} = f(x_n)$$

Thus, the fixed point x^* satisfies

$$x^* = f(x^*)$$

Now, assume that a small perturbation δx is added to x^*. The fixed point will be stable when, after application of the recurrence relation, the value of the perturbation decreases, and it will be unstable when the value of the perturbation increases or is kept constant under the recurrence relation. Application of $f(x)$ to $x^* + \delta x$ yields, up to the first order in δx

$$f(x^* + \delta x) = f(x^*) + (\partial f/\partial x)_{x=x^*}\delta x$$

The perturbation to x^* will decrease when

$$|(\partial f/\partial x)|_{x=x^*} < 1$$

In the situation described by (11.3), $f(x) = \beta x(1 - x)$, so that the value of $(\partial f/\partial x)$ evaluated at $x^* = (\beta - 1)/\beta$ is

$$|(\partial f/\partial x)|_{x=x^*} = |(1 - 2x^*)| = |2 - \beta|$$

Thus, when $\beta = 3$ the fixed point x^* is no longer stable. For high values of n, the values of x_n no longer tend to a fixed point x^*, but they take alternatively two values x_1^* and x_2^*, which satisfy the relations

$$x_1^* = \beta x_2^*(1 - x_2^*)$$

$$x_2^* = \beta x_1^*(1 - x_1^*)$$

The stability of the 2-cycle defined by x_1^* and x_2^* could be studied along the lines exposed in the preceding paragraph. It is found that it is stable for values of β that satisfy $3 \leqslant \beta < 3.449499$. At the latter value, a transition to a 4-cycle takes place. This is merely the beginning of an infinite sequence of bifurcations and period doublings. The first few values of β_k where a 2^k-cycle is born are

$\beta_1 = 3$ $\beta_3 = 3.544090\ldots$ $\beta_5 = 3.568759\ldots$

$\beta_2 = 3.449499\ldots$ $\beta_4 = 3.564407\ldots$ $\beta_6 = 3.569692\ldots$

When β exceeds the value $\beta_c = 3.569946\ldots$, the situation changes dramatically. Above this value, the trajectories take on values without following any recognizable rule, no matter how long we wait. In addition, two distinct situations, that is, two very close initial values $x_1(0)$ and $x_2(0)$, generate sequences of values which, once a certain time has passed, do not resemble each other at all. That is, the evolution of the system is such that apparently it "forgets" the initial conditions.

It should be noted that the interval $\beta_c \leqslant \beta \leqslant 4$ contains an infinite number of small windows of values of β for which there exists a stable m-cycle. The first such cycles to appear beyond β_c are of even period, but there are also odd cycles. For instance, the periodic 3-cycle first appears for $\beta = 3.828427$ and stays stable up to $\beta = 3.841499$. Outside these windows, there are no stable periodic orbits.

What relation exists between this example and Eqs. (11.1) and (11.2)? Certainly, if we calculate the ratio in (11.1) for the examples we are considering, in fact we obtain

$$\lim_{n \text{ large}} \frac{\beta_n - \beta_{n-1}}{\beta_{n+1} - \beta_n} \longrightarrow 4.669201$$

The same would happen with Eq. (11.2). It has been found that these numbers appear in numerous physical systems under diverse circumstances: for example, in the transition of laminar flow to turbulent flow in a fluid, or in electrical or chemical systems, and so on. These two Feigenbaum numbers appear universally and characterize the order–chaos transition; they seem to be fundamental for the description of certain aspects of nature. This reveals a certain regularity in chaos, or at least in one of the series of processes that leads to chaos.

11.3 STRANGE ATTRACTORS. MATHEMATICAL TREATMENT OF TEMPORAL CHAOS

In the mathematical example of the last section, we saw the appearance of an attractor. We can obtain a physical image for this concept if we consider the case of two blocks of metal that initially are at different temperatures. Once they are put in contact, after a certain time they reach an equilibrium state characterized by a uniform temperature. We say that the equilibrium is an attractor which has attracted the nonequilibrium situations toward equilibrium. There are other different attractors of a periodic nature called *limit cycles*, in which the object does not tend toward rest but rather oscillates about the equilibrium position. This oscillation can be very complex, but has a periodic or quasiperiodic character. There are other types of attractors, called *strange attractors*, which in the last few years have begun to change the concept of what science is, or at least the relationship between science and regularity.

In 1963, the meteorologist E. Lorenz proposed a set of equations whose impact has gone far beyond the field of atmospheric sciences that inspired them. The equations are

$$\begin{aligned} dx/dt &= -\sigma x + \sigma y \\ dy/dt &= -xy + rx - y \\ dz/dt &= xy - bz \end{aligned} \tag{11.4}$$

where σ, r, and b are positive parameters. These equations originated in the study of convection in the atmosphere. The surface of the Earth, warmed due to absorption of solar radiation, warms the atmosphere from below; from above, it loses energy toward outer space. The equations in (11.4) describe the convective movement by means of the variable x, and the horizontal (y) and vertical (z) variation in temperature by means of the variables y and z. The parameters σ, r, and b are respectively proportional to the Prandtl number (relation between the viscosity and the thermal diffusivity), the Rayleigh number (relation between the ascending convective tendency and gravity), and the size of the region whose behavior we are trying to reproduce. The complexity of the behavior of these equations is due to the nonlinear terms xy. In Fig. 11.2, we present the graphical solution of this system of equations with certain values of the parameters. In this case, the system oscillates irregularly between two situations, and its trajectory is extremely sensitive to the initial conditions.

In reality, Lorenz's discovery had been anticipated in some way by the mathematician H. Poincaré when he studied the three-body problem (for example, a planet rotating around two suns), although not explicitly.

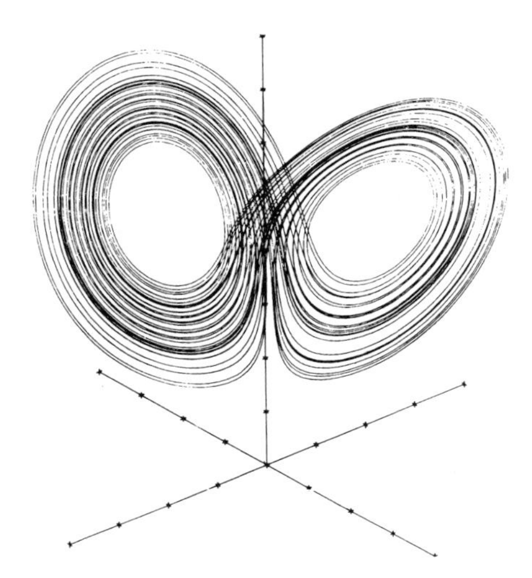

Figure 11.2 Lorenz attractor.

The work of Lorenz went unnoticed for fifteen years, until during the 1970s many scientists began to frenetically work on this type of problem.

Another model of chaotic behavior is the discovery by Hénon, generated by the recurrence relation

$$\begin{aligned} x_{n+1} &= y_n \\ y_{n+1} &= bx_n + ay_n - y_n^2 \end{aligned} \tag{11.5}$$

where the nonlinearity appears in the second equation (y_n^2).

What are the exceptional characteristics of these equations which have earned them the name *strange attractors*?

One characteristic of the solutions to these equations (obtained by calculations done on computers, the use of which has been fundamental for demonstrating this behavior) is that in certain cases, the projections of their trajectories rotate about two fixed points.

Thus, referring to Fig. (11.2), from a certain initial point a, the trajectory makes 15 returns around point 1, then 23 returns around point 2, then 73 returns around point 1, then 3 returns around point 2, and so on. The numerical sequence of the returns (15, 23, 73, 3, . . .) is rigorously prescribed by the deterministic equations (11.4) and the initial conditions. It is so difficult to solve the equations that for us it is practically unpredictable. On the other hand, the major surprise is the fact that, if we start from some initial conditions b which are very close to those which correspond to point a, the series of returns that give the solutions around the points (1, 2, 1, 2, 1, 2, 1, 2, . . .) is a numerical series completely different from the previous one. In addition, it has been proven that for any series of random numbers, there exists an initial condition that leads

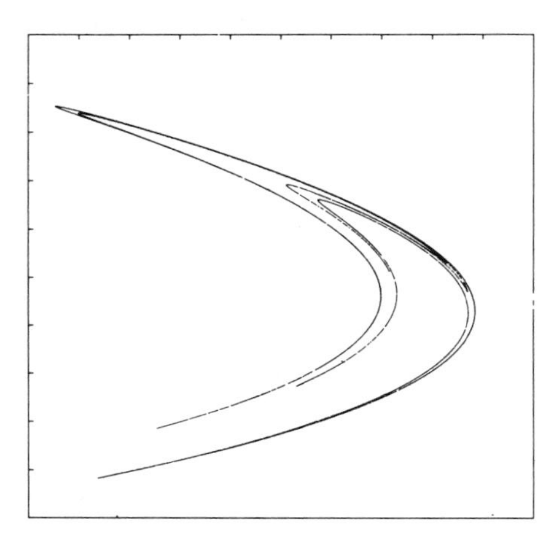

Figure 11.3 Hénon attractor.

to an evolution in which the number of returns of the trajectory exactly matches that series of numbers. This extreme sensitivity to the initial conditions and the chaotic aspect of the series of numbers allow us speak about deterministic chaos.

What physical examples of chaos can we offer? Let us start with meteorology. Points 1 and 2 in the last paragraph could be interpreted as zones of good weather or stable atmosphere (associated with high atmospheric pressures) and zones of bad weather or unstable atmosphere (associated with low atmospheric pressure). The sequence of the number of returns around points 1 and 2 is associated with the sequence of intervals of days of good weather or bad weather. The peculiar characteristics of a system of this sort would support the idea that it is impossible to know the evolution of stable or unstable weather with total precision, since the equations are of the strange attractor type and we do not know the initial conditions in complete detail.

Another example is the Earth's magnetism. The magnetic south pole does not always correspond to the geographic north pole. Over the history of the Earth, the position of the poles has inverted (about fifteen or twenty times in 11,000 years) with an irregular sequence in time. The stratigraphic study of the magnetism of rocks gives us good information about this inversion. Some physical models of the strange attractor type, based on the theory of magnetism and on the movement of terrestrial magma, allow us today to explain this phenomenon.

There are other examples in the field of physiology, such as ventricular fibrillation of the heart or the structure of the electroencephalograph waves in certain types of epileptic attack. And in hydrodynamics we find the best-known example, one of the most open problems in physics: turbulent motion of a fluid.

We especially want to emphasize the conceptual interest of strange attractors. Science has always been understood to be the explanation of regularities; strange attractors, on the other hand, involve irregularity. The discovery that much irregular behavior can be described in terms of very simple physicochemical equations has prompted, to an extraordinary degree, the study of this type of situation, which was previously considered pathological and due to a lack of experimental control.

11.4 FRACTAL DIMENSION. MATHEMATICAL TREATMENT OF GEOMETRIC CHAOS

In Chapter 10, we discussed order in space and order in time. In the last section, we saw how in some cases it is possible to treat temporal disorder mathematically. What are the possibilities for mathematical treatment of spatial chaos?

Richard Bentley, an English scholar who lived during the seventeenth century, described perfectly the motivation for fractal theory: "All pulchritude is relative. . . . We ought not . . . to believe that the banks of the ocean are really deformed, because they have not the form of a regular bulwark; nor that the mountains are out of shape, because they are not exact pyramids or cones; nor that the stars are unskillfully placed, because they are not all situated at uniform distance. These are not natural irregularities, but with respect to our fancies only; nor are they incommodious to the true uses of life and the designs of man's being on earth." [Cited by B. B. Mandelbrot in *The Fractal Geometry of Nature.*]

Noting the contrast between Euclidean geometry and nature is, then, an old idea. The objective of fractal geometry is to obtain a mathematical treatment of natural shapes that goes far beyond Euclidean geometry.

The Euclidean dimension in a space $\mathbf{R}^n$ coincides with the number of coordinates. We cannot satisfactorily understand the processes of irregularity and fragmentation with the definition of this type of dimension. We can distinguish between two types of dimensions. The Euclidean dimension D_T, which is an integer (in a space $\mathbf{R}^n$) between 0 and n; and the dimension D formulated by Hausdorff in 1919, which can be a real number between 0 and n. All Euclidean sets have $D = D_T$, while fractals, by definition, are those sets for which $D > D_T$.

A typical example involves the study of the profile of a coast. While from the topological viewpoint it has dimension 1, since it is topologically equivalent to a circumference, the fractal or Hausdorff dimension is different from one coast to another. In this case, what will be the fractal dimension?

The answer to this question in this example is easily comprehensible. In other examples, the answer can become complicated and even polemical. Let us then carry out the reasoning for the case of a coastline, and let us try to calculate its length. If we stop to think about it for a moment, we will conclude that it is not unequivocally defined. If we measure from kilometer to kilometer, its length will have a certain value. If we measure from meter to meter, the length increases tremendously, since we can penetrate the inlets, for example, while before we had at most counted the width of their mouths. If we measure it from millimeter to millimeter, we have a new spectactular increment in length (Fig. 11.4).

Thus the coastline is a nonrectifiable mathematical curve whose length depends on the unit of measurement. Despite the fact that from the topological viewpoint it is a line and, no matter how complex it is, it has dimension 1, according to Hausdorff–Besicovitch we define the dimension D as

$$1 - D \equiv \lim_{\epsilon \to 0} \frac{\log L(\epsilon)}{\log \epsilon} \tag{11.6}$$

where $L(\epsilon)$ is the length of the curve measured with the standard ϵ. This dimension indicates how the length increases as the measurement unit decreases ($L \approx \epsilon^{1-D}$, L increases when ϵ decreases). We can do the same

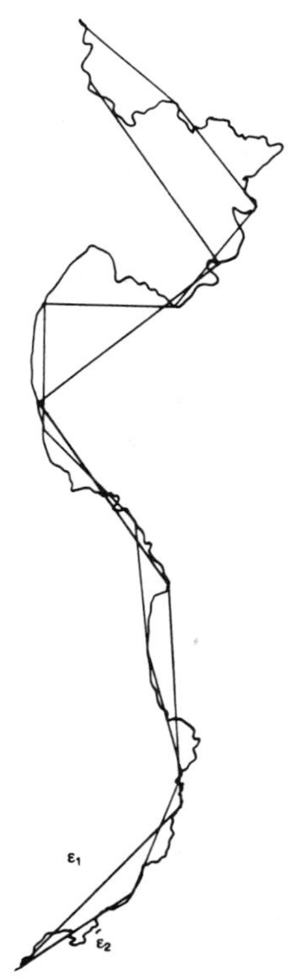

Figure 11.4 Length of the coastline measured with different standards $\epsilon_1 > \epsilon_2$; $L(\epsilon_2) > L(\epsilon_1)$

in the case of surfaces (usually considered of dimension 2) or for irregular distributions of points like the distribution of galaxies in the universe.

We can obtain an intuitive idea of the fractal dimension if we consider the erratic trajectory of a particle in a fluid undergoing Brownian motion, as we studied in Chapter 4. Despite having dimension 1, this curve covers almost the entire surface. In addition, when considering a small fragment of this line, a fragment which at first glance is rectilinear, we observe that in reality it is composed of an irregular line that is very similar to the total line. If we consider fragments that are increasingly smaller, this characteristic is maintained (that is, if we observe the particle every minute, every second, every thousandth of a second, and so on).

This allows us to intuitively understand that a curve with these characteristics has a dimension between 1 and 3, since despite being a line it practically fills the entire space. This latter statement is not true in the case of a rectifiable curve, since we always arrive at a scale on which its behavior is smooth.

Since 1980, scientific journals in all specialities have published hundreds of articles on the fractal dimension. Sometimes its calculation consists of a simple guessing game. Other times, it is necessary to go into more depth and to be able to correlate the fractal dimension of a curve with the physical laws of a phenomenon. For example, this is the case for fracture lines in a plate (glass, dry mud), for the irregularity of macromolecules in a solution, for the permeability of a porous medium as a function of the dimensions of its pores, for the irregular distribution of galaxies in the universe, for the profile of petroleum deposits as a function of the porosity of the ground, and so forth.

Before giving an example of this type of calculation, we mention again, albeit briefly, the example discussed in Chapter 3 referring to the bronchial tree. This is a good example of the diversified domain of application of the fractal concept.

When we observe the bronchial tree, two apparently contradictory facts leap immediately into view: the great variability of the dimensions of the bronchioles (diameters, lengths) and the high degree of organization. Despite its asymmetries, an ordering principle exists in the "design" of the bronchial tree. As in the erratic trajectory of a particle in a fluid, there is no unique measurement scale. If we observe the structure of the bronchioles in greater detail, then new levels of detail appear. This characteristic property of fractals is called *self-similarity*, and it is this property that gives such an organized appearance to the bronchial tree.

In Chapter 3, using the principle of minimum entropy production we obtain an expression for the diameter of the bronchioles as a function of the generation z. The result, as is mentioned in that chapter, is valid up to the twelfth generation. For higher generations, the discrepancies with respect to the experimental results are appreciable. However, the result itself is in principle not very congruent with the existence of multiple scales, since the very fact that a formula is obtained which gives the diameter of a bronchiole as a function of its generation indicates existence of a unique scale. B. J. West and A. L. Goldberger, analyzing these discrepancies and incongruencies and given the self-similarity of the bronchial tree, proposed the use of a fractal model for the bronchioles whose most notable characteristic is the incorporation of multiple scales. Using the fractal model and group renormalization methods (the theory used with notable results in many areas of physics), they obtained appreciable reconciliation with the experimental results, no matter what the value of z. In addition, the model shows better adaptability when confronting changing conditions, which is useful for studying the morphogenesis of the lung, for example.

11.5 FRACTAL DIMENSION OF PROTEINS

As an illustration of the fractal dimension, let us choose an example of biological interest: the fractal dimension of the tertiary structure of proteins.

The structure of proteins has been successfully analyzed by X-rays. Regular, distinct types of conformation have been revealed in their structure which are apparent after careful analysis of their tertiary structure. This is very complicated, and can be described only qualitatively.

Obviously, it is necessary to obtain an analytical method for protein conformation which will be quantitative and objective, and which will extract information about the irregularity of the structures of the molecules. Until recently, the most common tools were Euclidean geometry and differential geometry. These techniques are limited to the study of shapes like the circle, the ellipse, the parabola, the sphere, and differentiable curves or surfaces. For extremely irregular shapes like those we deal with here, or like the path followed by a Brownian particle (Fig. 11.5), these geometries do not work.

This is the context in which fractal theory is applied. The theory deals with irregular shapes and provides quantitative means for extracting regularity from an apparently irregular shape. In this way, let us illustrate how we can determine the fractal dimension of the tertiary structure of proteins.

Let us define the length of the protein L as a function of the measurement standard ϵ [the same parameter ϵ that appears in expression (11.6)], starting from the following definition (see Fig. 11.6).

Starting from the atom C^{α} of the N terminal residue, we draw a zigzag line that connects the C^{α} atoms of the protein at intervals from ϵ

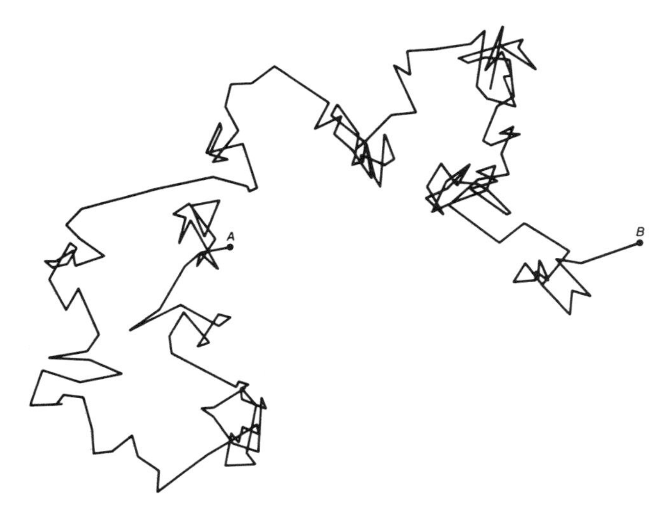

Figure 11.5 Erratic behavior of Brownian particles.

to ϵ residues. If, when we arrive at the last section of the molecule, a quantity less than ϵ residues remains, we stop the process at this point.

As in the example of the coastline in the last section, we note that the smaller ϵ is, the greater is the level of detail with which we observe the molecule.

The length of the protein $L(\epsilon)$ is calculated as the sum of the lengths of the different sections of the zigzag line plus the contribution of the final section, which can be evaluated on the basis of the number of residues, the size ϵ of the scale, and average length of the different sections of the zigzag line.

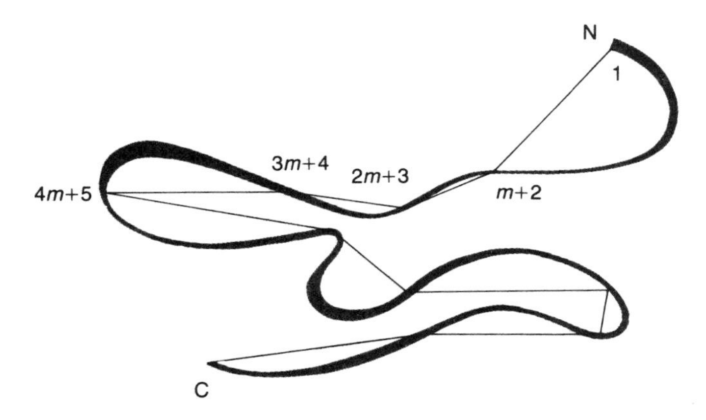

Figure 11.6 Measurement of the length of a protein.

In this way we can construct $L(\epsilon)$ as a function of ϵ, starting from the information contained in the reference data banks for proteins.

Representing log $L(\epsilon)$ as a function of log ϵ, we obtain a rectilinear zone for $1 \leqslant \epsilon \leqslant 10$, and for $\epsilon \geqslant 10$ we obtain a zone that is less regular but adjustable to a straight line of greater slope than in the previous case. In Table 11.1, we present some concrete values for the fractal dimensions (the slopes of the lines mentioned) for some proteins with diverse secondary structures. On the average, we find that the fractal dimension corresponding to short distances ($1 \leqslant \epsilon \leqslant 10$) is 1.34, and for long distances ($\epsilon \geqslant 10$) we obtain 1.95. These values can be compared with the fractal dimension that we would obtain for a Gaussian chain (a chain composed of identical rectilinear elements, one after another and oriented at random, without attraction or repulsion between its elements), which is 1.5. We see that, for short distances, the fractal dimension is less than the dimension of the Gaussian chain ($1.34 < 1.50$), and for long distances it is greater ($1.95 > 1.50$).

The corresponding interpretation is that, at short distances, the molecule is more extended than a random chain, due to steric repulsion be-

TABLE 11.1 Fractal Dimensions of Different Proteins

Secondary structure	Protein	Number of residues	Fractal dimension: Short distance	Fractal dimension: Long distance
α helix	Hemoglobin	141	1.40	2.15
	Myoglobin (sperm whale)	153	1.42	2.01
β sheet	Immunoglobulin	208	1.26	1.67
	Trypsin (pH 8)	223	1.30	2.14
α helix and β sheet in separate zones	Lysozyme (egg white from hen)	129	1.42	1.90
	Ribonuclease A	124	1.33	1.90
α helix and β sheet in alternating zones	Adenylate kinase (pig)	194	1.36	1.80
	Phosphoglycerate kinase	408	1.33	1.97

Source: Y. Isogai and T. Itoh, 1984.

tween nearby atoms. But at long distances, the molecule tends more to contract into a compact agglomeration than would occur with the Gaussian chain, due to attractive forces between residues.

We see, then, how the idea of repulsive interaction at short distances and attractive interaction at long distances translates into a difference between the fractal coefficients at small and large distances, compared with the case without interactions. More detailed studies, beyond the range of this introduction, can lead to more specific conclusions of greater interest.

In the discussion up to this point, we have again met a situation similar to the one we confronted when dealing with strange attractors. In the last ten years, we have learned to discover a certain order (or at least, a certain classification) in the domain of complexity, and in some cases we have been able to relate these general ideas to well-known and apparently regular laws of classical physics. The latter has thus gained a new conceptual thrust.

BIBLIOGRAPHY

CRUTCHFIELD, J. P., J. D. FARMER, N. H. PACKARD, and R. S. SHAW, "Chaos," *Scientific American* 255, no. 6 (December 1986), 46–57.

HOLDEN, A. V., *Chaos*. Manchester: Manchester University Press, 1986.

ISOGAI, Y., and T. ITOH, "Fractal analysis of tertiary structure of protein molecule," *Journal of the Physical Society of Japan* 53, no. 6 (1984), 2162–71.

MANDELBROT, B. B., *The Fractal Geometry of Nature*. New York: Freeman, 1977.

RANADA, A. F., "Movimiento cáotico [Chaotic motion]," *Investigación y Ciencia*, March 1986, pp. 12–23.

WEST, B. J., and A. L. GOLDBERGER, "Physiology in fractal dimensions," *American Scientist* 75 (1987), 354–365.

Index

C

D

E

F

G

H

I

J

K

L

M